Contents

End

END

Cosmic Catastrophe and the Fate of the Universe

FRANK CLOSE

SIMON & SCHUSTER
LONDON · SYDNEY · NEW YORK · TOKYO · TORONTO

First published in Great Britain by
Simon & Schuster Ltd in 1988

Simon & Schuster Ltd
West Garden Place
Kendal Street
London W2 2AQ

Simon & Schuster of Australia Pty Ltd
Sydney

British Library Cataloguing-in-Publication Data

Close, F.E (Frank E)
End: cosmic catastrophe and the fate of
the universe.
1. Universe. Evolution – Forecasts
I. Title
523.1'1

ISBN 0–671–65461–6

Typeset in Palatino 11/14pt by Inforum Ltd, Portsmouth
Printed and bound in Great Britain by
Richard Clay Ltd, Bungay, Suffolk

Acknowledgements

Many individuals and organisations have been most helpful in providing material, advice, free education and comments on early versions of all or parts of the manuscript. Some of these are anonymous, such as people who have asked me a question or made some comment after talks that I have given; the originators of several anecdotes that I have accumulated are unknown to me; and my many discussions with speakers at the British Association for the Advancement of Science, at Rutherford Appleton Laboratory and at the European Centre for Nuclear Research in Geneva (CERN) are too numerous to mention other than collectively.

I would also like to express some personal thanks. Professor V. Levin from the USSR took time out from a conference in Hungary to tell me of his experiences in the Tunguska region of Siberia and provided much of the colour for that part of the story. The library staff at Rutherford Appleton Laboratory and at CERN have helped me trace obscure literature on many occasions. I would like to thank Nick Brealey, Gill Close, Paul Davies, Ali Graham, Mike Green, Beverly Halstead, Iain Nicholson, Liz Paton, Roger Phillips, Martin Rees and Sheila Watson for their reading of parts or all of various versions and Ted Barnes, John Barrow, Victor Clube, Brian Combridge, Nigel Henbest, Duncan Ollson-Steel, Subir Sarkar and Peter Worsley for their help in clarifying some specific questions. I think that I adopted around 90 per cent of your suggestions: I hope that this final version meets with your approval. Thanks to you all.

Frank Close
Oxford
September 1987

For the girls who enjoyed Jupiter's moons
and can read the sign

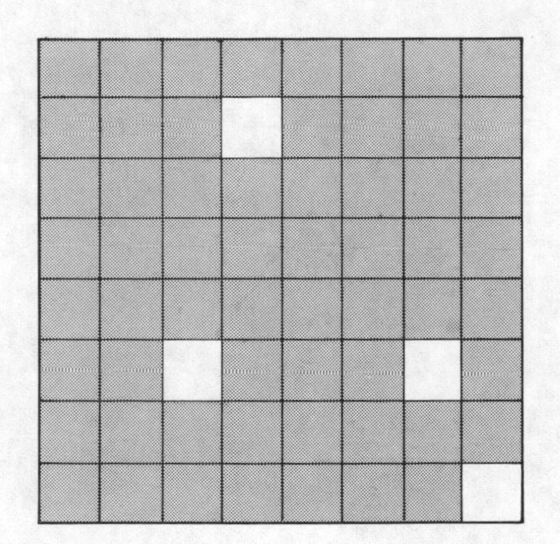

1

Apocalypse - When?

Beyond Xanadu, in the Temple of Brahmin, priests are counting the days to the end of time.

According to legend, the temple contains three divine pyramids of stones representing Brahma, the Creator, Vishnu, the Preserver, and Shiva, the Destroyer. On the first day Brahma created The World and built a single pyramid of 42 stones, the largest at the bottom.

Each day at sunset the priests move one stone from Brahma to Vishnu, Vishnu to Shiva or perhaps directly from Brahma to Shiva. The only rule is that they must never put a large stone on top of a smaller one. Eventually all stones but one will be moved from Creator through Preserver to the pile belonging to the Destroyer. As the Sun sinks and the final stone is moved, the priests' work will be completed. Brahma created and now Vishnu destroys. The Sun will never rise again.

If the legend were true, how long will the Universe last – how many moves must they make?

It is easy to see that, if there had been only two stones, a total of three moves would have done the job. With three stones, the number more than doubles to seven moves. The solution is shown in Figure 1.1. You may care to discover the sequence for four stones – it takes 15 moves. Each extra stone more than doubles the number of moves required. For 42 stones the priests must make one move less than two times two, 42 times ($2^{42} - 1$). At one move per sunset this will take them a little more than 10 billion years, which is how old the Universe is – NOW!

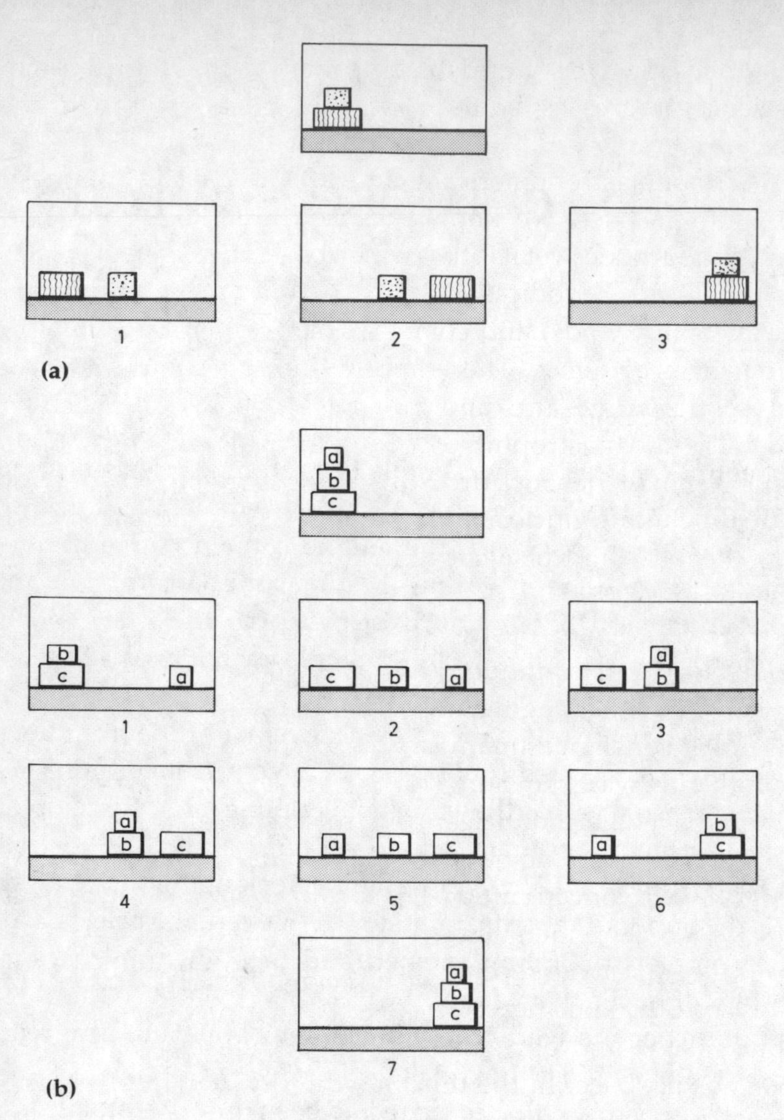

1.1 The end of the Universe puzzle

The Universe may indeed be this old but the Sun and Earth are much younger; slightly less than 5 billion years old in fact. The priests are only half way through their task, so the Sun and we still have 5 billion years to go – and modern science agrees with this estimate of the Sun's future.

One day 5 billion years hence the Sun will be gone and we shall be too. However, worlds end in many ways, and the more that science progresses, so the more contenders emerge for the Apocalypse.

Though astronomers are familiar with constellations and even remote galaxies, we know relatively little about our immediate neighbourhood, yet it is that which poses the greatest threats in the short term. Periodically, lumps of rock, even pieces of minor planets (asteroids) and comets, drop in from the sky – the scars remain as evidence. There have been suggestions that the extinction of the dinosaurs and the ends of geological epochs may be witness to catastrophic collisions in the past. How long before another maverick rock drops in, or one big enough to destroy a vast area including nuclear power plants? The Chernobyl disaster in 1986 gave a hint of what that additional hazard could do to the environment.

If we are not destroyed by extraterrestrial invaders, or by ourselves, our continued existence will require the Sun to go on steady and unchanging, much as it appears now. In reality the Sun is not as constant as many people think. We are now learning a lot about the way the Sun works, and can even peer inside and examine its thermonuclear core; some strange things appear to be going on that we do not yet understand. In the second part of the story, I will concentrate on our relationship with the Sun, describe how well we understand it and how we are trying to resolve the puzzles that have recently been discovered.

The Sun is the nearest star to us but it is only one of a billion in our galaxy, and our galaxy in turn is but one of countless such islands in space. In the third part of the story I will describe how those more distant stars can affect us.

As we swing round the Sun, so is the Sun dragging us on a grand tour of the Milky Way. Not everywhere is as bland as the part we are currently in. We will meet clouds of dust that can affect the climatic balance; we might encroach too near other stars, which could disturb our orbit, moving us out of the narrow zone that makes life comfortable or even flipping us out of the solar system entirely.

Not all stars are as stable as our Sun. Stars can explode – known as 'supernova'. In 1987 one lit up as bright as a whole galaxy in the southern skies. The cataclysm took place nearly 170,000 years ago but so far away that news has only just arrived. It was too distant to cause us harm, but the blast from a nearby supernova could rip off our atmosphere.

In Part IV we will look at the Universe as a whole and evaluate its future. By re-creating the sort of conditions present in the first moments of the Big Bang, we are learning about the start of the Universe and gaining clues as to how it may all end. Will it expand and cool or instead collapse and fry us?

We will also look into the make-up of matter – the stuff that everything in the known universe appears to be made of. Imagine the ultimate erosion where everything in sight, including the atoms of your body, instantly collapsed or changed into some new form. Bizarre though this may sound, recent discoveries in particle physics suggest that it might be possible. We are also getting hints that there may be 'strange matter', more stable than us, which could destroy us, the Earth, and everything if it came here in enough concentration. There are even serious suggestions that the structure of space itself may be unstable, and that the entire universe could 'slip downhill' if some trigger unfortunately occurred somewhere.

These ideas are very new; we could not have imagined them seriously even ten years ago. At the very least they are making us re-examine our place in the universe and question our assumptions that it will always be as it is now.

We are living at a critical period in history. Once primitive peoples feared storms, the night, and lived by superstitions. Then science rationalised things, created order and brought us to the point where we can create theories of Genesis and test them in the laboratory. We began to feel omnipotent. We were aware that there are man-made threats – nuclear weapons, stocks of deadly germs – which could wipe us from the face of the Earth. But the Universe – that would go on for ever.

Now we are not so sure. We are becoming increasingly aware of our vulnerability and so far have done very little about it. With planning we could one day escape the Earth and colonise space –

after all, transatlantic flight is commonplace today but would have been regarded as dreaming in Columbus' time

Some scientists believe in the anthropic principle – mankind's arrival is so improbable that it is as if Nature conspired to bring it about. They see hints that the Universe created life to be its agents for immortality. As far as we know we are the carriers of that task. If we can avoid extinction in the short term, be it self-inflicted or from external causes, then we may propagate throughout space into the indefinite future. 'We' may bear little resemblance to the human flesh and bone of today: in a mere billion years primitive molecules have evolved into us who can contemplate the Universe within which we live; what will another billion billion bring?

You and I have no right to life; we inherited it by chance. Now that we are here we have the duty to do our bit in the great human relay race. If the Universe is to avoid ultimate extinction, then life – and that may mean us here on Earth – has to keep going. If we can learn how to live with the possibility of nuclear destruction we then have to deal with the threat of natural extinctions. I conclude with some ideas that are emerging on how we might deal with them and on what the long-term future of life might be.

To advertise that this isn't idle musing, I will begin with an example of a near catastrophe that happened this century.

Part 1

Our Own Backyard

2

Cosmic Close Encounters

Beyond the Ural Mountains, 1,000 miles East of Moscow, lies a vast unspoiled region of swamps, rivers and forests. Stretching from the Arctic Sea in the north to Mongolia in the south, and from the Urals to Manchuria, it is a sparsely inhabited area larger than the whole of Western Europe, known to only a handful of outsiders. There are few roads, fewer towns and for much of the year everything is covered by a blanket of snow. Strange things have happened here and many years passed before the rest of the World learned of them.

In the remote heart of this lonely continent is the hidden valley of the Tunguska river. Here for a few weeks during the summer the snows recede and reindeer graze among the endless pine trees. Into this peaceful scene, one morning in June 1908, a comet burst from outer space. In an instant the reindeer and the trees were annihilated for 30 miles around. A 10 million tonne gravel-filled snowball, with a diameter longer than a football field, exploded in the atmosphere with the force of several hydrogen bombs and sent shock waves around the whole World. It threw so much dust into the stratosphere that sunlight was scattered from the bright side of the globe right around into the Earth's shadow. A quarter of the way round the World in London, 6,000 miles away, the midnight sky was as light as early evening. Everyone realised that something odd had happened, but what and where?

Even today, 80 years after the event, very few outsiders have visited the devastated site. To see the scars you have to make a major expedition. You must choose your starting time carefully so as to reach the site, and get back again, during the summer. Set out from Moscow at the end of April on an internal flight eastwards. As the plane rises shakily into the sky you look down with a comet's eye view of the Earth. Only occasional houses are scattered around and soon even these give way to continuous forest. The ribbon of the Trans-Siberian railway is the only clue that people live on this planet.

We reach central Asia and must leave the relative comfort of Aeroflot and transfer to a small plane for the journey over the mountains. The going is tough. Your pilot may have to wait for up to two weeks for the weather to clear enough to fly at all. Even then the nearest airstrip is still many miles from the site and you must continue on foot. Arrange with your pilot for food to be dropped by parachute at places on the route. White water rushes through gorges and canyons; trees cling to the sheer rock faces. Crossing the terrain will take three weeks if the summer is dry, but, if the spring has been wet, the rivers will run high and your inward journey (let alone your return) may take as long as two months. Even the flat areas aren't easy. There are extensive swamps and persistent mosquitoes. At last you come to a tree-filled valley – the valley of the Tunguska river. The Tunguska is named after the Tungus people, a small ethnic group who survive by hunting bears and deer in the forests. It was they who saw and later recounted the fateful day in 1908.

The first scientist to reach here was the Czechoslovakian Leonid Kulik in 1927. At the point immediately below the explosion he saw a vast mud plain as if a thousand bulldozers had cleared the forest to prepare the foundations for a city the size of London. Surrounding this bleak scene was a ring of charred tree stumps. Beyond this the trees lay scattered like matchsticks, felled by a tumultuous hurricane, the blastwave from the exploding comet. Life had been totally destroyed, and remained so for over a quarter of a century.

The lonely desolation of the site matches the awful privations suffered in reaching it. It is this very remoteness that has hidden it from the World, leaving us cosily ignorant of the forces of cosmic assault. This was not the first time that the Earth was blitzed from outer space, nor is it likely to be the last. It is merely the most recent.

Pictures from spacecraft show the blue jewel of planet Earth brilliant against a black velvet of seemingly empty space. This image of a solid Earth rushing through an empty void is deceptive. On our journey round the Sun – a journey of half a billion miles in 30 million seconds – we are accompanied by over half a dozen other planets, various moons, asteroids, comets, volcanic dust, gas, nuclear radiation, neutrinos, the solar wind and other bizarre bits and pieces; all of them hell bent on their own journeys, their paths interweaving and criss-crossing. Every second we rush through 20 miles of space and if anything is there already waiting for us, or is heading towards the same spot as us, then we hit it.

The Earth is constantly under fire. About 1,000 tonnes of extraterrestrial debris hit the upper atmosphere each day. Most of the pieces are so small that the atmosphere burns them to cinders, leaving the familiar trails known as shooting stars. But on timescales of millions of years much larger objects can hit us, and will continue to do so once in a while.

Things are on the move all around us and the signs are there to see. Go outdoors on a fine night and look up at the sky. A man-made satellite comes into view: the rays of the sun catch it and make it visible, shining for a few minutes as it orbits around the globe. A shooting star suddenly streaks leaving a trail across the sky; no sooner have you realised that it is there than it is gone. Both of these sights are quite likely if you look at the sky for an hour on a clear night away from city lights. Occasionally a comet may appear, like Halley's comet in 1986, and be visible to the naked eye for weeks on end. If you live in the far north, you may be fortunate enough to see a display of cosmic rain – such as the aurora borealis or northern lights.

Man-made Hazards

There are several thousand man-made satellites on record, and many unlisted military ones, all orbiting in the sky. In addition there is a lot of junk – extinct rocket motors, astronauts' spanners and pieces that have fallen off satellites. The thin atmosphere gradually brakes them and eventually they fall, usually burning up in the process. Occasionally something goes wrong and a satellite makes an unexpected premature return. On such occasions the world's news media have a good story for a day or two as scientists continually update the projected landing site and public relations spokesmen blithely assure us that the satellite's nuclear power source poses no real threat. Eventually the thing plunges into some remote place. The laws of chance make it very unlikely that something dropping from the sky at random will end up landing on a city. As the journey to the remote Tungusku reminds us, overpopulated though the World may appear, the uninhabited *area* vastly exceeds the populated.

These man-made hazards are occasional annoyances and embarrassments, but rarely more than that. They are vastly outnumbered by the natural junk that is continuously hitting the Earth.

Natural Junk

Smallest of all are pieces of atoms that rain down on the upper atmosphere above our heads.

These are produced by violent processes deep in space, such as when stars explode. Powerful forces eject them into space where some, approaching too near to the Earth, are entrapped by our planet's magnetic poles, and pulled in.

Scientists sent balloons into the upper atmosphere to meet these cosmic rays and recorded images of them. They gave some of the first hints of the power within the nucleus, and much of modern nuclear science has grown from those early discoveries. Their power is immense – a single atomic particle in the cosmic beam may have enough energy to raise a human 3 cm off the ground.

When these rays hit the upper atmosphere their energy dissipates as they disrupt atoms in the air and produce a shower of less powerful subatomic particles. These finally reach ground as a gentle rain – interesting for scientists but no real hazard for humans, although long-term exposure at high altitudes or in high-flying aircraft rapidly increases the chance of skin cancers developing from this radiation.

The auroras are seen only by people in extreme northern or southern latitudes, near to the Earth's magnetic poles. Even if you have not seen this spectacle, at some time or other you will probably have seen the results of rather larger extraterrestrials crashing into the upper atmosphere.

Meteors are pieces of dust from dead comets out in space. Each time a comet swings in towards the Sun it loses some of its ice. Gradually the cement holding the gravel all melts away and a shrapnel of stones and boulders fly independently around the solar system like the rings of Saturn on a larger scale. The small stones spread out all around the orbit and form a long sausage of debris. When the Earth passes through one of these we experience a meteor shower. We rush towards the individual stones at 20 miles a second, and gravity pulls them Earthwards, wind friction burning them to white heat. It is this that causes the bright trail – the 'shooting star'.

You can see one or two on any clear night as we hit random pieces in space. We pass through one of the Sun's 'rings' at the same point on our annual orbit and on those nights you may see dozens of meteors each hour. For example, every August we pass through one of these tubes of debris and the result is a meteor shower called the Perseids. There are other rings of debris circling the Sun and when we meet them, regularly each year, our atmosphere burns some up and produces a meteor shower.

Occasionally, however, some quite sizeable pieces survive and make landfall. We call these meteorites.

Meteorites: Pennies from Heaven

Stones falling from the heavens have been recorded in the literature for millennia. Whereas comets usually stay up in the sky, visible to all, 'stars' falling from the heavens are seen by only a few people but can be quite terrifying. They are the ultimate in *son et lumière*.

First a fiery mass appears in the sky – what we see is the

compressed air in front that heats up by the friction of the rock's motion and can be far bigger than the rock itself. The surface of the rock melts, sparks fly off the tail and smoke trails remain long after landfall. Then it may disintegrate and unseen dark pieces fall to ground at 200 metres per second – the speed of a jet in a nose dive. There is a shock wave that rumbles and echoes 'like the sound of guns thundering in battle'.

One morning in 1972, a meteorite shot through the skies in the Rocky Mountains, bright enough to show up in broad daylight. The scenery in those parts is exquisitely beautiful and provides an ideal background to Nature's display of awesome power, reminding us of the primaeval forces that fashioned the planet. Keep on the lookout – these fireballs are rare but not so rare as you might think. On the average one or two fireballs like this occur somewhere on Earth each week. Between 10 and 20 less brilliant ones also arrive each day.

Meteorites may be either stone or lumps of iron. In general they have a different chemical and mineral content to Earthly rocks; as a result it is easy to recognise one on the ground even if no one saw it actually fall. Laboratory tests have shown common features suggesting that some meteorites may be fragments of a single large body, possibly the size of Earth, that has broken up, its remnants orbiting the Sun in perpetuity. One theory is that they, and some of the asteroids, came as a result of a planet disintegrating after a collision with some other large body. If this is true then it hardly lessens our concern that some day the Earth could be similarly devastated.

Iron meteorites can withstand the shock of hitting the atmosphere whereas stone meteorites tend to break up. If the explosion takes place high up, then the shower can be huge. 100,000 stones fell in one shower during 1868 in Poland. 10,000 fell in Holbrook, Arizona, in 1912, and many thousands in a shower in the Soviet Union during 1947. In a large shower most pieces are smaller than grape shot. Lots of dust lands and can show up as dark powder if it falls on a snow-covered landscape. The largest known stone meteorites fell in a shower of 100 over Kansas in 1948, and include one monster of 1 tonne. There are other large ones on record including one of over half a tonne in Long Island, New York, one third of a tonne in Finland and similar weight in Czechoslovakia.

These must be the large fragments of even bigger rocks that hit the outer shield of air, tens of miles above our heads. Indeed they are puny compared to the largest pieces of iron that have fallen. The biggest of all that is still visible on the surface weighs 60 tonnes, the weight of a dozen juggernauts, and lies where it fell on a farm in South-West Africa. From debris around the area it is estimated that this is the largest fragment of an iron core that weighed 100 tonnes. These pieces are the largest on the ground but there is no record of their arrival in the distant past. In the Soviet shower of 1948 a single piece weighing 2 tonnes was seen actually falling to Earth with astonishing fury.

Meteorites hit the upper atmosphere at speeds of up to 50 miles per second and then air resistance slows them down. The damage that they do depends upon the amount of kinetic or motional energy that they have. Two objects moving at the same speed as each other have energies in proportion to their masses: if one is twice as massive then it has twice as much energy as the other. So if a small stone is travelling at the same speed as a car, the stone will have only a millionth as much energy as the car and will do correspondingly less damage in a collision.

The damage also depends on how fast they are travelling. If you double the speed then you quadruple the energy; treble the speed and the energy increases ninefold. So a fast-moving stone can have as much energy as a slow-moving car. In fact, a piece of dust weighing only 1/10 gram and moving at 50 miles in a second has as much energy as a 1 tonne car travelling at 50 miles an hour! A small stone weighing 1 gram hits the atmosphere with a punch similar to that impacted by a fast-moving truck. The ultimate monsters enter at Mach 50. These are the ones that disappear far into the ground and leave a deep extensive crater as a scar on the landscape.

Down at sea level we are protected by the Earth's atmosphere, but out in space even small bits of gravel could be lethal. The energy released if one hit a spacecraft could be a thousand times that of an equal weight of TNT. A pinhead particle could make an airleak in the hull; a pebble the size of a fingertip could destroy the spaceship entirely. There have even been suggestions that the occasional and tragic break-up of airliners at altitude might be due to meteorite impact shattering a tailfin or other crucial equipment.

The awesome sights and sounds announcing the arrival of a meteorite from the heavens convinced primitive peoples that the gods had sent them. As a result meteorites became treasures and were venerated in temples as literally 'gifts from the gods'. In the Acts of the Apostles, chapter 19, v 35, we read of the Ephesians who worshipped the goddess Diana and 'the symbol of her which fell from heaven'. King Maximilian I of Germany went off to the crusades encouraged by a stone falling near Ensisheim in the Alsace during 1492. The stone still resides in the village church.

When you see 'relics of the cross' you are rightly sceptical. Similar scepticism attended claims to possess 'stones that fell from heaven' in the eighteenth century. The emerging acceptance of science, led by the French Academy of Sciences, decreed that such irregular phenomena were impossible. Claims to have examples of meteorites became discredited as relics of a superstitious bygone age and many European museums threw them out from their exhibits.

Pennyweights from heaven started to become respectable after an iron meteorite fell in Austria in 1751. Several hundreds of people heard the thunder and saw the ball of fire in the sky. Stories about the fearsome vision of the night spread far and wide and pieces of iron were found scattered around. With speculation rife about this supernatural experience, the church quickly became interested and priests set out to interview witnesses. Clerics collected sworn testimonies and sent them to the Austrian emperor. In turn this stimulated E. F. Chladni, a German physicist, to defend the reality of the phenomenon and scientists began to take an interest. Meteorites became 'official' when a shower of stones fell near Paris in 1803. Even the French Academy of Sceptics could not ignore a shower in their own backyard and at last the whole scientific community agreed that meteorites really do fall from the sky.

The idea of cosmic catastrophes is very popular in science fiction and disaster movies. However, unlike some of the bizarre offerings in that genre, rocks dropping from the sky and destroying whole cities are not impossible. There is plenty of evidence for gargantuan invaders. Remote sensing satellites send back images of the ground showing that the Earth is pitted with the evidence of impacts.

Known impact craters that exceed 1 kilometre in size fairly pepper the globe, as we can see on the map (Figure 2.1). There are more than a hundred of them and these are only a small percentage of the total. The vast regions of central Asia, Africa and Brazil, where few craters are listed, have been explored less than other areas. Where meteorites have fallen in the jungles, or in geologically active regions, all traces will have been overgrown or otherwise obliterated by the living Earth. Oceans cover most of the surface and anything landing here will leave no trace (though one suggestion is that the Caribbean Sea is the result of a huge impact several millions of years ago). Only in dry undisturbed areas like the Arizona desert or the cratered arid regions of central Australia do the marks survive in pristine form.

In the south-western corner of the United States lie thousands of square miles of beautiful desert wilderness. The air is so still and the sun so bright that you can see a grey windswept plain stretching for miles against the backdrop of a turquoise sky. In the middle of this open country is a grotesque scar, a huge hole left one day 30,000 years ago when a lump of rock the size of an oil-tanker dropped from the sky and hit the ground while travelling thirty times faster than Concorde. The heat of the impact vaporised the soil and threw up a huge cloud high into the stratosphere. Boulders bigger than houses rained down from the cloud, adding to the devastation.

The crater remains, uneroded, much as it did once the dust had settled all those years ago. It is so huge – over 1 mile wide and 3 miles in circumference – that it is easily visible from outer space, like a piece of the Man in the Moon. Photos taken from satellites 900 miles high show the Colorado river, the extensive desert and a single pockmark where the boulder landed.

The crater may appear insignificant from the spy satellite, but come down to Earth and see it in close-up (p.1 of photos). The City

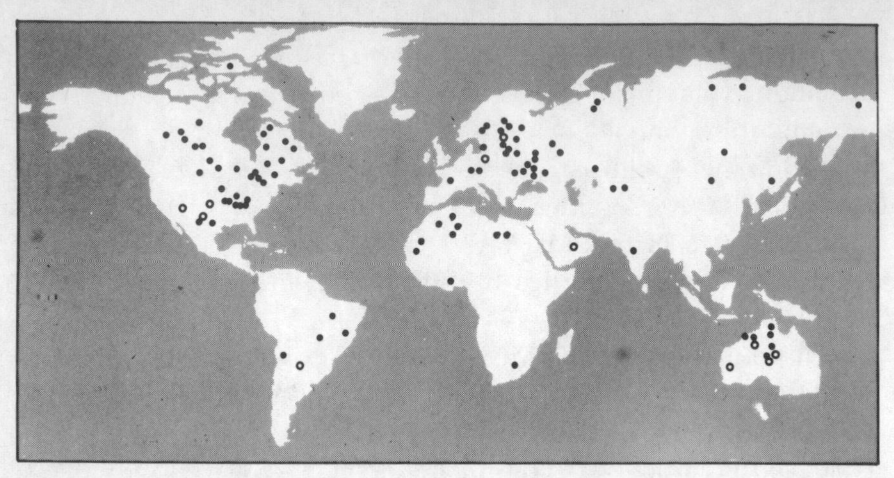

2.1 Terrestrial crater sites Open circles denote craters less than 1 km across with meteorite fragments and shock features. Dots denote larger older structures (from R. Grieve, 1983).

of London could be placed inside it and very few of the buildings would even reach the rim.

Hordes of tourists come and marvel at this awesome hole in the desert. Many of them think that it is unique. The Tungusku event already shows us that it is not. We are not alone in space and unwelcome visitors drop in from time to time: woe betide anyone who is around when the next one calls.

You might wonder what is the largest impact that the Earth has survived, or might be likely to suffer. Although the ravages of time have covered over most of the traces of impacts on Earth, the dead worlds elsewhere in the solar system preserve a record of collisions. These give us clues of what Nature has on offer. The Moon is our nearest neighbour and its scarred face shows what the Earth would be like if there were no atmosphere to protect us. We can see huge craters on the Moon from 250,000 miles away. Imagine them at ground level and how they were fashioned.

The Moon's craters vary in size, many being smaller than a coin while half a dozen exceed 100 miles diameter, as big as Sicily or the length of Long Island. Many of these were caused when the Moon was very young and the planets were still being formed. The cooling pieces of small 'protoplanets' hit the Moon and left their marks for ever. However, these are not really relevant for estimating the likelihood of collisions on Earth *today*. What we need is evidence that major collisions have occurred in recent times and are still going on.

When the Apollo astronauts visited the Moon they left behind four seismometers to measure Moonquakes. When these relayed signals back to Earth, the listening scientists were surprised by the sounds of large meteorites hitting the Moon, some as big as 10 metres across.

The flux of hits of the Moon varied through the year, peaking with the known meteor showers. The biggest one is at the end of June when we pass through the Taurid meteor stream. We on Earth didn't notice much thanks to the blanket of air surrounding us, but the airless Moon carries the scars. The stones hitting the moon sandblasted the old craters, creating the moondust that the astronauts kicked up as they walked the surface.

The Moon is a good detector and is teaching us a lot about what is going on 'out there'. From the distribution of hits – their size and frequency – we can estimate the number in the swarm (we only sample a small fraction), and it comes out close to 1 million. We can also calculate the distribution of sizes and so estimate the size of the biggest. If there are 10,000 of 1 metre and 1,000 of 10 metres diameter, then we can be fairly sure that there will be many of 100 metres and a few of 1 kilometre and maybe one of tens of kilometres. Now 'one' could be zero or it could be two or three – the statistics are risky at the end of the distribution. But there is little doubt about the rest: there will be several whose girths are larger than a football field. We know what such impacts can do: Arizona's crater and the Tunguska event are examples.

There are many pieces in the annual Taurid meteor shower that we would be aware of if we ran into them. In the last few years we have found objects in this stream that are 10 miles across. Such impacts would threaten life on Earth but fortunately are exceedingly rare. There is a greater chance of encountering a swarm of Tunguska-sized objects.

With modern technology we ought to be able to contemplate doing something about these. But how long do we have to prepare, and how big an invader can we expect? Surprisingly little work has been done – we are adept at looking deep into the Universe but the timetable of debris in our own backyard is not very well known.

For the moment we take refuge in the game of chance. The Moon's craters provide a record over the aeons, which suggests that a major impact, leading to a crater exceeding 1 mile, should occur once every 10,000 years on average; the larger they are, the rarer.

Think of what lies within a mile of your home and imagine it all extinguished. Even half a mile is not trifling and impacts of that size should happen much more often. The chance that one will happen in a populated area is small, as the Earth's area is dominantly uninhabited. The Tunguska event is a good example; people around the world noticed its occurrence like the Krakatoa eruption, but it didn't affect many people directly. Had the invader been bigger it could have disrupted much of the atmosphere. The extinction of the dinosaurs 60 million years ago may have been an example of such an extremis.

The Universe is a hostile place and there is nowhere to hide from its effects. Arizona yesterday, Siberia today; where next?

3

The Neighbourhood

Mercury, Venus, Earth, Mars, Jupiter, Saturn, Uranus, Neptune and Pluto. Nine planets are orbiting the Sun in regular processions. In addition, billions of tonnes of cosmic debris are swarming around. Some have rained down on the innermost planets scarring them for life. Debris is still pouring in. Comets zoom in from deep space; rocks drop out of the sky and burn up as meteors; some asteroids have orbits that cross our own.

The Sun outweighs all the planets put together and its gravity holds the planets, comets and asteroids in continuous orbits on their recurring journeys.

The innermost planets are like the Earth – small, with surfaces of solid rock enclosing cores of molten iron. Then come four giants consisting of hydrogen, helium, ammonia and methane all frozen solid in cold depths of space. Jupiter weighs more than the rest put together, its gravity so strong that it has a miniature orbiting system of its own with over a dozen known moons. Saturn and Uranus also have entrapped many moons and rings of smaller rocks. Neptune awaits the Voyager probe in 1989 and we expect to find a complex system of satellites there too. Beyond the big four lies tiny Pluto, whose satellite, Charon, is almost as big as it is.

Although Jupiter outweighs the rest, it is still 1,000 times lighter than the Sun; it is the *Sun*'s huge gravity that dominates the solar system.

Isaac Newton proposed his law of gravitation in 1687. He showed not only that apples drop on to people's heads but that the planets are held in orbits around the Sun too. We know we're safe

3.1 Conic sections A horizontal section gives a circle, an inclined cut gives an ellipse. If the Sun is at the focus of an ellipse (either of the dots) planets will orbit along elliptical paths. Cut the cone parallel to one side and you get a parabola. If the Sun is at the focus, the path is the line that goes off to the left for ever – such as a trapped comet that makes a single pass but that is always ensnared by the Sun's gravity. Cut the cone vertically and you have a hyperbola. This is the path that a body follows if it is not trapped by the Sun but is merely deflected by the solar gravity in its passage.

from them because Newton showed what sort of paths solar orbiters can follow. Newton's law implies that if a body is attracted to a massive object (such as the Sun) with a force that weakens in proportion to the square of the distance, then the body will move along one or other of a set of characteristic paths. These are the 'conic sections', the shapes that you get when you cut ice-cream cones (see Figure 3.1).

The Earth and the major planets all follow orbits that are nearly circular and keep well away from one another. Even Neptune and Pluto, whose orbits *do* cross, are never at the crossover at the same time. Some of the asteroids follow pronounced ellipses that cross our orbit twice.

The Sun is at the focus of the ellipse (see Figure 3.2), quite unlike the portrayal on the British pound note which showed it at the centre. It is amusing to recall that Newton was once in charge of Britain's Royal Mint. His successor in the 1970s, responsible for the

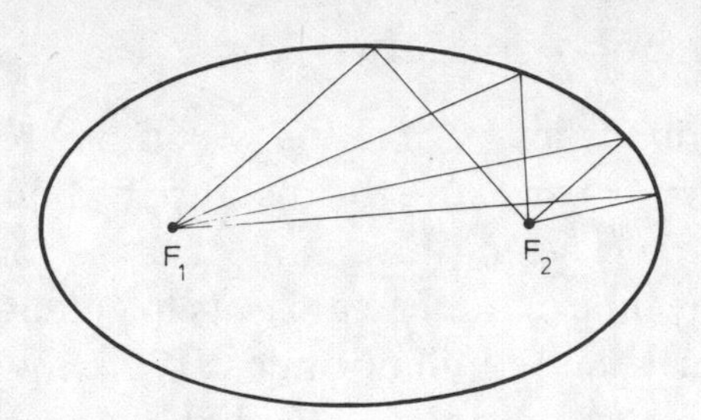

3.2 Ellipses The sun is at the focus of the ellipse. There are two foci, F_1 and F_2. The distance from F_1 to F_2 via the edge is always the same. To make an ellipse choose a piece of string pinned to the paper at two points near enough that the string is slack. Now stretch it tight with a pencil and, keeping it tight, draw a curve. The pins will be at the foci and the curve will be an ellipse.

design, was obviously not educated in Newton's theory of gravity and the orbits of planets around the Sun.

Periodic comets, such as Halley's, follow very elongated elliptical orbits. From its period of 76 years, and Newton's law of gravity, we can calculate the orbit and find that its remotest point is further from the Sun than is Neptune. It hurls in, as in 1985/6, crossing our orbit once on the way in as in November 1985, once on the way out as in April 1986, missing us both times to return again another day.

It is 300 years since people realised that comets could be travelling around the sun in very elongated ellipses or making a single pass on a hyperbola. In the former case they would be part of the solar system, much as the planets are, but coming in from far far away and being visible only for the short time that they are near the sun. In the latter they would be cosmic wanderers that happen by chance to be deflected by the sun's gravity en route from the distant past to a lonely future. Of the 700 or so known comets some 600 have periodic orbits of the elliptical kind, the remainder being once-only voyagers.

We will look at these mavericks that cross our orbit and put us at risk in Chapter 4. Here we will concentrate on the nearly circular orbiters. They pose no threat but are interesting in that they provide fossil relics of past collisions showing what has happened in the past and could happen again.

Measuring the Solar System

Our nearest neighbour in space is the Moon. Astronauts can get there in a week. We can measure the distance by bouncing radar or laser beams off it and seeing how long it takes for the signal to return – it takes the time of three heartbeats to get there and back, a round trip of nearly half a million miles. We know the speed of light, or radar beams, so accurately that we can measure the distance of the Moon to a precision better than the thickness of this book. We can even tell that it is receding from us at a rate of about 3 cm each year – 3 metres in a century.

So not only are things on the move around one another, but their orbits are changing. The Moon raises tides in the seas and strains the rocks of Earth; as a result of these tidal forces it is receding and has become locked in to us so that it always presents the same face. The Earth too in its journeying round the Sun is gradually slowing its spin; in some distant future it will present the same face continuously to the Sun. Atomic clocks can record this slowing – a fraction of a second each year – and every so often we have to adjust our time clocks by adding a second on to a year: 'time stands still' for that moment. Conversely, in the past the Earth spun faster and so there were more days per year. Prehistoric geological records show that there were 400 days per annum several hundred million years ago. So over the epochs everything is readjusting – a theme that will pervade our story.

The radar beam that measured the distance from Earth to the Moon can also reach to Venus. Our nearest planetary neighbour is some 30 million miles away at its closest approach.

The next benchmark in scaling the Universe is the distance to the Sun. We can't bounce radar beams back from the Sun so we have to start with the information on the distance to Venus and use that to set the scale. Newton's theory of gravity implies that the time taken to orbit the Sun depends on the distance you are away from it, not on your mass. (As we can predict the occurrence of eclipses to within seconds we have no doubt that the theory is reliable.) The further away you are, so the slower you move (or else you would fly out of the solar system like a car trying to take a corner too fast), and as you have further to travel in a circuit it takes you longer. If you are four times further away than another body it will take you eight times as long (in general, the time taken is in proportion to the distance multiplied by the square root of the distance, hence four times two equals eight in our example).

Now the Venusian year is 224 of our days, which means that on average the distance from us to Venus at closest approach is about one-third of the distance from us to the Sun. The radar blip showed us how far away Venus is from us and so we can work out how far away the Sun is. We travel around the Sun in an elliptical path so the distance isn't the same throughout the year but the mean distance is 93 million miles. If you could fly to the Sun on Concorde it would take about a dozen years. A ray of light makes it in just over 8 minutes.

Bode's Law

The nearly circular paths of the planets have sizes that fit with a simple numerical rule. We're going to compare the distances from the Sun to the planets with that from Sun to the Earth. For simplicity we will define the Earth to be 10 units distance from the Sun. Then Mercury, Venus, Mars, Jupiter and Saturn are very nearly 4, 7, 15, 52 and 95 units distance respectively.

These were the only planets known in 1766 when Johann Titius first noticed that these numbers fit into a series. If you subtract 4 from each then the data are very similar to a series where each successive number is double the previous one: 0, 3, 6, 12, 24, 48, 96, 192 and so on. On adding four in again you have 4, 7, 10, 16, 28, 52, 100, 196. These are very good comparisons with the orbits out to Mars. The member at 28 is missing but at 52 Jupiter occurs on cue and Saturn is only 5 per cent out.

Johann Bode succeeded Titius at the Berlin Observatory, published Titius's rule and got the credit. This numerology is often referred to as Bode's law.

No one paid much attention to it until 1781 when Caroline and William Herschel discovered Uranus. Its distance from the Sun is 192 in these same units. This is quite remarkable since the Titius–Bode rule predicts 196. The agreement was sensational enough

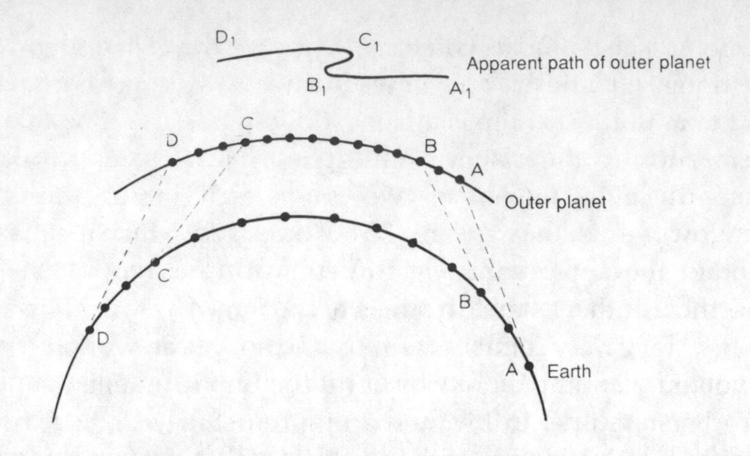

3.3 Retrograde motion As the fast-moving Earth overtakes a slower-moving more remote planet, the planet's motion appears to change direction in the sky. When the Earth and planet are at respective positions A the line of sight to the planet is the line joining the two A, which is projected on the distant heavens at A_1. The points B, C, D and the lines joining them show the respective lines of sight at later times. Between A and B the Earth is coming round the corner and the planet appears to be overtaking us (moving leftwards across the sky). From B to C the rapid Earth overtakes the planet and causes its apparent motion to reverse.

that astronomers began to wonder about the missing '28'.

A group of astronomers led by Baron von Zach set out to look for a missing planet but were beaten by Giuseppe Piazzi, director of the Palermo observatory in Sicily. On 1 January 1801 – 1.1.01, an easy date to remember – he was looking at the constellation Taurus when he saw a small unknown star. It moved – it could not be a star! He followed it for a whole month, during which time its motion reversed (an effect due to the Earth overtaking a body that is orbiting the sun more remotely than we – Figure 3.3). He named it Ceres, after the patron saint of Sicily.

Ceres is about one-fifth the size of the Moon. Its distance from the Sun agrees with the Titius–Bode rule but its orbit is inclined at 10 degrees to the plane in which the other planets revolve. So it isn't quite like the other planets. Moreover it is very small. People weren't entirely convinced that this was the whole story and so they kept searching.

In March 1802. H. Olbers discovered Pallas, another asteroid whose size and orbit are very like those of Ceres. Two more soon followed – Juno and Vesta, with sizes about one-quarter that of Ceres. These moderate-sized objects, all at the same mean distance from the Sun and fitting collectively with Bode's missing '28', suggested that they are the remnants of a larger body that exploded in the distant past, or are planetessimals that did not assemble into a planet in the first place.

By 1890, 300 had been found. It was a laborious job and involved scanning the sky by eye looking for anomalies among the known starscape. In 1891 photographic searches began with telescopes rotating in time with the Earth so that the fixed stars appear as points on the image. Planets and asteroids show up as streaks as they move slowly across the sky. Within the first half of this century 2,000 were found this way. Not a year passes without several new ones being discovered. There is no sign that the discovery rate is falling off, which suggests that there are several hundred more still to be found. And in the last weeks of 1986 scientists rediscovered one that had been lost track of decades before.

It was the search for a planet 2.8 times more remote than the Earth from the Sun that led to the discovery of the asteroids. Most, but not all, of them are whirling around between Mars and Jupiter. The total mass of all of them adds up to less than 1/2000 of the Earth. Ceres alone accounts for half of this. Collisions among the asteroids, one with another in this congestion, have probably broken several of them up. Many have orbits that are elliptical rather than circular and several cross our path. Some may have hit us in the past.

Collisions in the Solar System

Between 1962 and 1975 the United States launched a series of spacecraft known as the Mariner probes. They gave the first close-up views of Mars, Venus and Mercury, and, in the airless conditions of the planet Mercury, the evidence was found for the near destruction of an entire planet. Mariner sent back pictures of the Caloris Basin, a crater that is nearly 1,000 miles across. Whereas the Arizona meteor crater is a spot in the desert, the Caloris Basin would fill Arizona along with much of the six states covering America's south-western corner; almost a quarter of the whole sub-continent! The shock waves were so strong that they travelled right through the planet, leaving gorges and hills on the far side.

In 1977, the United States launched two further probes, the Voyager spacecraft, which have given us close-up views of the outer planets and their environs. So far they have visited Jupiter (1979), Saturn (1980) and Uranus (1986). We already knew that Saturn has beautiful rings and Voyager showed that Jupiter and Uranus have too. These three giant planets have many moons and the Voyager pictures reveal that these moons contain a record of aeons of impacts. The messages imprinted on the surfaces of these moons are making some scientists on Earth ponder. These lonely worlds bear testimony to the power of Nature. They have gathered it and kept it for millennia, unknown until the day that the first space probe arrived.

Voyager's pictures of Callisto, one of Jupiter's many moons, showed the results of violent impacts in unexpected detail. Callisto is more than 3,000 miles across, nearly half the diameter of the Earth, and its entire surface is utterly cratered, not a single square metre has escaped. Indeed, you could not create a new crater without destroying one that is already there. Rocks and huge boulders must have been crashing into Callisto for millions of years.

Callisto's surface is a mixture of ice and rock, a global Antarctica. Glaciers flow into the meteor craters. The biggest impact sites were originally marked out by huge holes and high cliffs, but have now levelled out so that only the surrounding walls remain. The smaller craters survive in their pristine form; the ice can support the wall of a small crater permanently but the larger craters collapse.

Among this awful desolation is a spectacular and beautiful bull's-eye of multiple rings. A stupendous collision, singular even among Callisto's continual bombardment, made a crater 400 miles across. This melted the ice beneath the surface and huge blast waves of water spread outwards. The temperature is 180 degrees below zero. In this biting cold the waves froze into mountainous rings 2,000 miles around. We know that this happened rather recently because there are no craters on its floor; this bulls-eye has splattered the earlier craters and in time will be peppered again by further ones.

This is an apt illustration of the raw power of Nature and reminds us of our puny insignificance. Who could have imagined the reality of frozen mountains the extent of the USA, centred on a crater larger than the state of Kansas? While this gives us pause, hopefully it is not a good guide to what we are likely to feel on the Earth. Callisto is near to Jupiter, whose huge gravity pulls debris into its stratosphere. So its moons are peppered with flying rocks much more often than they would be were they near to the Earth. Callisto inhabits the solar freeway whereas Earth is off the main road.

After visiting Jupiter's environs Voyager set off deeper into space reaching Saturn in 1980 and giving our first views of its cratered moons.

One of Saturn's moons, Mimas, is only 250 miles in diameter but contains a crater that is 7 miles deep and a full 60 miles across. A crater that is one-quarter the size of a moon represents a collision in extremis; a slightly larger impact would probably have shattered Mimas. Indeed, it appears that Mimas may be flawed right through, as there are cracks on the far side opposite the crater.

This is reminiscent of Mercury's cleavage and shows that that is not unique. Nor is this the limit of Nature's potential devastation; even more violent collisions have happened to the Saturnian moons – one of which *has* been broken in two. The two halves are still circling and are the moons with the unromantic names 'S_{10}' and 'S_{11}'.

These twins orbit Saturn at a distance of 100,000 miles yet their paths lie within 30 miles of each other, little more than the width of the English Channel, closer than the northern and southern extremities of New York City. S_{10} is 150 miles across while S_{11} is an odd shape, a fragment that is 50 miles high and 20 miles across; twice as long in one direction as the other. The only way that Nature could have fashioned such irregular shapes is by breaking up a larger object.

These co-orbiting moons are not on *exactly* the same paths and so they take slightly different times to orbit Saturn. Consequently they must pass each other periodically. It is a mystery how they manage to do this since their sizes are bigger than the separation of their orbits. But now astronomers believe that, instead of rotating periodically, some objects in the solar system behave chaotically. Jack Wisdon of the Massachusetts Institute of Technology showed in 1987 that Hyperion, one of Saturn's moons, is now tumbling chaotically. Orbital chaos was once thought to be a dynamical impossibility but astronomers now think it answers a variety of questions about the solar system. Chaotic orbits may explain how asteroids reach Earth in the form of meteors. They add a new and utterly unpredictable ingredient in the risk stakes.

Voyager spent the next five years, from 1980 to 1985, travelling onwards from Saturn to Uranus. At last, on 24 January 1986, it sped past Uranus and found again evidence of craters and even shattered satellites. It is possible that Uranus itself has been flipped over by some ancient collision. Whereas the Earth's magnetic poles are near to its rotation axis, the magnetic poles of Uranus are nearly on its equator; unique in the solar system.

Voyager also found more evidence of chaos. Miranda, one of Uranus' moons, 300 miles in diameter, has a peculiar geology, with rifts and oval gorges, which appear to be the result of it having been tumbling and lurching erratically for millions of years in chaotic motion instead of rotating at predictable intervals.

At the farthest outposts of the planetary solar system, where Voyager will reach some day, lie Pluto and Neptune. (Voyager will meet Neptune, but Pluto is currently on the far side of the Sun and so won't be encountered this time.) We tend to think that Pluto is the farthest away from the Sun, but in fact these two planets criss-cross in their orbits, and currently Neptune is the farther away. (This will continue to be the case until March 1999 when Pluto will cross over and return to the outside.) There is no danger of these two colliding, however. Their orbits take 165 and 248 years, neatly in the ratio of 2 to 3. Pluto goes round the Sun twice while Neptune goes round three times. When Neptune is at the crossover point, Pluto is somewhere else. When Pluto arrives at the crossover, Neptune has moved on to another place. So it continues, round and round. These two planets are safe from one another.

Percy Lowell, the great American astronomer, initiated a search for a planet beyond Uranus and Neptune. He died in 1916 and it was not until 1930 that Pluto was found. The discovery was announced on Lowell's birthday and the symbol for the planet is ♇, a monogram of his initials and the first two letters of the planet's name.

Pluto is smaller even than our own moon. Furthermore it is only three times heavier than its own moon, Charon (discovered by James Christy in 1978). It is hard in the circumstances to think of Pluto as a planet, rather it and Charon form a pair of planets orbiting one another as they move collectively around the Sun.

Pluto is an overgrown snowball and may even be a satellite that escaped from a parent planet, probably Neptune. There are several pieces of circumstantial evidence supporting the idea that at some time in the past a maverick body passed near to Neptune and disrupted its moons. Pluto may be one of them, ejected outwards and now looping in and out of Neptune's path as if trying to orbit that planet as they both slowly circle the distant central Sun. This is unique in the whole solar system – all the other planets have orbits far removed from one another.

Another of Neptune's moons, Triton, behaves strangely. All other close-in moons orbit their parent planets in the same direction as our moon orbits the Earth; Triton orbits Neptune the opposite way. Our moon is also typical in following the equatorial regions, pursuing the Sun through the heavens; Triton's path is inclined at 20 degrees to Neptune's equator – it is as if our moon looped up over our heads and dipped below the horizon as we faced the sun.

Finally there is Neptune's other moon – Nereid. This moves so fast that Neptune can barely hold on to it. And it doesn't keep its distance, it comes in close and then moves off far away on a tight ellipse. Imagine our moon sometimes filling the sky close in and then heading off to a remote small disc and then hurtling back in again. That is how Nereid would appear to a Neptunian.

In 1979 R. Harrington and T. van Fladern wrote an article in the astronomy journal *Icarus* explaining how all these orbits might have arisen. We can even test their theory because it predicts that a tenth planet exists far beyond Pluto.

Suppose that Neptune originally had four moons in near circular orbits, much like Jupiter's largest moons. All that is required is that a planet, three times the bulk of Earth, plunges through the satellite system. The cannibal planet captures the innermost moon and carries it off into deep space. The second nearest moon also escapes and ends up in a remote orbit: Pluto. The third moon has its orbit flipped and is Triton, while the fourth just manages to survive in orbit around Neptune and is Nereid.

This is not contrived. If you start off with four reasonable circular orbits and study the effects of a bulky object passing through, then these weird orbits emerge naturally. And that is indeed what Pluto, Triton and Nereid look like today. The search is now on for the lost planet, the tenth planet – Planet X. Perhaps in the outer reaches of the solar system we have the most extreme examples of what can happen when encounters get too close. Had such near misses involved the Earth, they would have been terminal for mankind.

Jupiter: King of the Planets

Astrologers are much enthused by planetary alignments. Although most people do not take astrology very seriously there are still many who think that there might be a physical basis for its predictions, especially when all the planets are in the same part of the sky. For example, in 1982 many people believed that the gravitational forces engendered by a 'grand alignment' of the planets would act in concert and could provide a physical mechanism for triggering earthquakes or other disasters. Indeed, one mystic even predicted World's End.

Although at first sight this may appear very reasonable, in fact the sums simply do not add up. Gravity is the only force that measurably affects the motion of planets and their moons. The force of gravity acting between two bodies is in proportion to their masses and varies inversely as the square of the distance between them. In other words, it is bigger the more massive the bodies and feebler the further apart they are. We are trapped by the Sun because it is the most massive thing around here, outweighing the rest of the solar system 500 times over. The Moon, by contrast, is lightweight but it is very near by: it is its nearness that makes it play an important role. Everything else is insubstantial and remote; for example Jupiter, the biggest planet, has less than 0.5 per cent of the Sun's mass and is at least five times further from us.

The Moon's motion around the Earth raises tides as it orbits. Gravity's strength weakens with distance, so the tug on the Pacific Ocean is greater when the Moon is directly over Hawaii than when it is away round the corner above the Atlantic.

Similarly the revolution of the planets raise tides in the Sun, causing small bulges in their wake. The outer planets orbit slowly while the inner ones fairly rush round. Once in a while the planets will all be on the same side of the Sun and all tug in concert. Devotees of the so-called 'Jupiter effect' noticed that between 1977 and 1982 the planets would be grouped on the same side of the Sun. They believed that the collective gravitational tug would stretch out the Sun, causing great disturbances on its surface.

Now it is indeed the case that we on Earth are much more affected by inconstant behaviour in the Sun than perhaps we realise. The Sun may appear to be a shining ball far away from us, but its thin non-luminous outer atmosphere extends far beyond the Earth. We are literally journeying *inside* the Sun! Storms in the glowing Sun can reach its outer regions and directly disturb our own atmosphere, interfering with radio communications and affecting the weather. According to the advocates of the 'Jupiter effect', if the Sun is really disturbed, then it would severely disrupt our upper atmosphere and perturb our rotation. The jolt would strain the Earth's crust, causing disastrous earthquakes as weak spots gave way.

These dramatic catastrophes sound plausible because of the imagery in the emotive phrase – 'stretching out the Sun'. However, the planets do no such thing; their effect is nugatory. The Sun is oscillating up and down by several miles all the time without us being particularly aware of it (the glowing Sun is nearly 1 *million* miles across so these oscillations are comparable in relative scale to the tides on the Earth). The additional effect due to the planets aligning amounts to less than the thickness of this book. So the planetary tides in the Sun are over a million times less effective than the lunar-induced tides on Earth.

Changes in the Sun, like all random motions, are the result of forces acting – which means gravity in the present context. A common misconception is that the motion of the planets around their common fulcrum is important. Proponents argue that when the giant outer planets are all lined up, the centre of the Sun will be furthest away from the centre of the solar system and so the Sun will 'pull in the opposite direction to balance the added effect of the planets'.

However, the fulcrum plays no role in determining the size of the forces acting on the Sun or anywhere else. It is important to remember this before getting too carried away with predictions of disasters stemming from *distant* planets. The outer reaches of the solar system are incredibly remote. The distances from the Sun roughly double with each successive planet (recall the Bode numbers on page 25); thus a journey from the sun to Jupiter, passing Mercury, Venus, Earth, Mars and the asteroid belt, brings you only half way to Saturn. When you reach Saturn you are still only half way to Uranus, and Uranus is only the midpoint in a trip to Neptune (see Figure 3.4).

As if this isn't enough to make the outer planets impotent, the tidal forces result from the gravitational force *difference* across the body that is of interest and die away in proportion to the distance *cubed*: double the distance and the effects die eightfold. The result is that little close-in Mercury raises tides in the Sun almost as effectively as distant mighty Jupiter does and Venus is comparable to Jupiter. Jupiter is less important than Mercury, Venus and the Earth combined.

None the less, Jupiter could cause us problems *indirectly*. As comets swing in from the cold depths of space, they may pass near the huge outer planets and be ensnared by their pull. Comets that would otherwise have looped in around the Sun and off again to

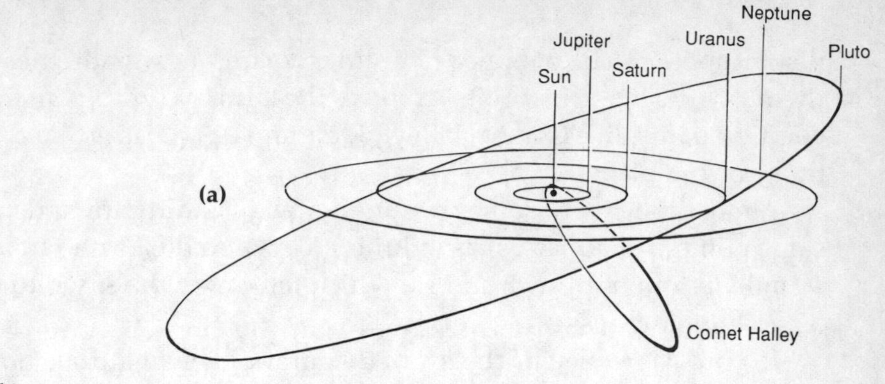

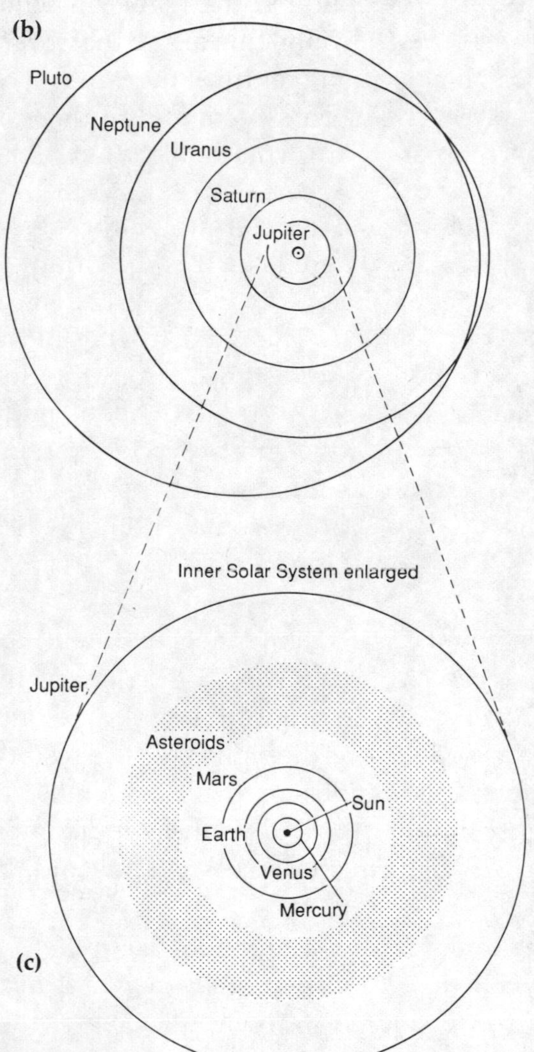

3.4 The solar system
(a) All of the known planets, with the exception of Pluto, orbit the Sun in the same plane. Pluto's orbit, and that of Comet Halley, are very different from this. This is one reason for suspecting that Pluto's orbit has been drastically disturbed in the past.

(b) Projected onto a plane we see that Pluto's orbit crosses Neptune's. The relative distances to the outer planets are so vast that the inner solar system has to be shown on an enlarged scale (Fig. 3.4c).

distant space can be whipped like slingshot on a new path. Most of them are trapped in orbits around the Sun, whirling in tight elliptical paths, like Comet Halley. Some end up in paths that cross those of the planets.

Almost inevitably some day one will be heading towards the same point in space as us, as in June 1908 when the Earth ran into 10 million tonnes of rock and ice which fell out of the sky into the Siberian wilderness.

If you were placing bets on the most likely natural global disaster, collision with an asteroid or a large piece of a dead comet are leading contenders.

4

Chicken on the Roundabout

Comets

On 30 August 1979 a satellite orbiting the earth recorded a catastrophic event – the death of a heavenly body. The Sun's gravity had ensnared Comet 'Howard-Koomen-Michels 1979XI', which fell headlong into the stellar furnace. Within a matter of seconds the comet, larger than the Earth, was destroyed. The immense power of the Sun vaporised the comet as an elephant destroys an ant. The debris scattered for millions of miles in the Sun's atmosphere.

Comets date back to the start of the solar system and are probably its most primitive members. They are balls of gravel and ice, most of them spending much of their time far beyond Pluto in deep space where we're unaware of them until one zooms into view on a looping path towards the Sun, whipping behind it and setting off back into deep space.

The comet's head is deep frozen when far from the Sun where the ambient temperature is −270 degrees Centigrade. The ice contains deep-frozen ammonia, methane and amino acids, the stuff of life. There are suggestions that cometary impacts may have contributed to Earth's early terrestrial atmosphere and brought the primitive organic molecules for prebiotic evolution.

As comets approach the Sun, its warmth vaporises the ice. The ejected gas and dust reflect sunlight and appear from the Earth as a bright head, or coma. The comet's nucleus consists of one or two lumps of rock of about a mile diameter. The coma is usually bigger than the Earth and may be as much as 100,000 miles across. The solar wind (high-speed particles coming from the Sun) and the radiation drive very fine dust particles and ionised gas from the coma, forming a lengthy tail which always points away from the Sun. This can extend for enormous distances, even as long as the distance from the Earth to the Sun. This elongated shape in the heavens is the characteristic signature of a comet, recorded in the Bayeux tapestry, paintings and literature.

The melting ice releases the stony pieces with the result that the comet eventually becomes a mile-sized ball of gravel weighing over a million billion tonnes. Large though this sounds, it is trifling on the scale of the Earth and other heavenly bodies. As far as we know, comets have never perturbed the motions of the planets or even their moons. In 1770 and again in 1886 comets passed between the moons of Jupiter without noticeably affecting the moons' motions even though the comets' trajectories were significantly deflected. So a near miss will not affect the course of the Earth; it will take more than a comet to send us headlong into the Sun or flip us out into deep space.

A direct hit is a more serious matter. Comets are moving at several miles per second, as is the Earth, and a direct hit from a mile-sized lump of rock can't be ignored. The Arizona and Tunguska impacts involved objects 'only' 100 metres across.

Some of the comets pass near the major planets, Jupiter and Saturn, and are dragged in. Small pieces break away leaving a trail behind the main mass. The beautiful rings around Saturn are probably pieces of dead comets, as are the rings around Jupiter, unknown before the Voyager spacecraft arrived there. And in January 1986 Voyager arrived at Uranus showing us rings there too. Cometary fragments are everywhere.

There are huge rings around the Sun. When the Earth passes through one of these we experience a meteor shower as the dust burns up in our atmosphere. Some of the larger pieces reach ground as meteorites. The largest pieces may orbit on long after the rest have dispersed or been burned up in collisions. These extinct comet heads form some of the asteroids.

The largest asteroids are as big as countries. Ceres spans France and Belgium. Pallas and Vesta are as big as Southern Scandinavia and Juno is the size of Ireland (see Figure 4.1). None of these comes anywhere near our orbit to threaten us, but many much smaller asteroids actually cross the Earth's orbit. These mavericks are called 'Apollo Objects' after the half-mile diameter asteroid Apollo, which was discovered when it approached near to the Earth in 1932.

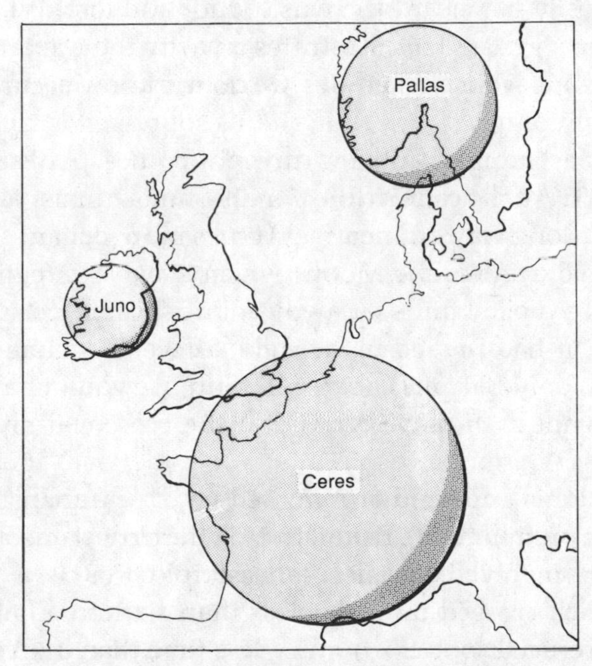

4.1 Asteroids Asteroid sizes compared with European countries. Ceres, the largest, obliterates France; Pallas covers southern Norway and Sweden; Juno compares to Ireland.

Apollo Objects

Recent searches for Earth-crossing asteroids suggest that they could be numerous. None of them is larger than a few miles in diameter but a collision would wreak havoc for hundreds of miles and the disturbance of the atmosphere would cause great storms at least.

The first hints of such an awful possibility emerged in 1898 with the discovery of Eros, an asteroid that comes within Mars's orbit.

For the first time astronomers realised that asteroids are not restricted to the space between Jupiter and Mars. Eros takes only a few months longer to orbit the Sun than we do. In doing so it encroaches nearer to us than we ever get to Venus, our nearest neighbour among the planets.

Soon other examples of wandering asteroids turned up. In 1911 a small one cosily named Albert was found, and lost. Where it is now is anyone's guess. This illustrates a worry – there are Earth-crossing asteroids whose schedules we do not know accurately or at all.

Gradually as more examples turned up, the perils became sharper. In 1932 Apollo came within 7 million miles of us. Now that might sound a long way off, nearly 30 times more distant than the Moon. However, unlike the Moon, which stays at a regular distance from us, Apollo whirls far away and rushes in, crossing our path. Indeed, it had missed us by only six weeks. If the Earth's orbit had been only 100 miles nearer the Sun, we would have been moving fast enough to have arrived at the crossover six weeks earlier.

In 1936 Adonis appeared and missed us by less than 12 days. The insurance premiums are rising! In 1937 the closest encounter in modern times involved Hermes, an asteroid that is a mile in diameter, which crossed our orbit less than six hours before we arrived at the crucial spot. Six hours – less time than it takes to fly the Atlantic.

There's no sign that these occurrences are unusual. On 20 October 1976 we were missed by less than half a day. And if you feel confident that 'surely by now we must know the timetable and there can't be any new shocks to come', I offer you the night of 28 February 1982 when two 'Earth-grazers' were discovered.

H. Schuster at an observatory in Chile discovered an asteroid that comes near to our orbit but never crosses us. Later that night E. Helin at the Palomar Observatory in California caught sight of an asteroid that was alarmingly close. Tracking its path from night to night soon showed that it was flying away from us and had already crossed our orbit twice! On its way in towards the Sun it had crossed our path three months earlier and had done so again on its way out just three weeks before Ms Helin sighted it. No one had detected it while it was heading towards our orbit. This highlights one of the major problems: these Apollo objects are large enough to damage us severely but they are too small to detect easily.

In 1987 the asteroid Icarus passed within 4 million miles of us – a mere 16 times more remote than the Moon. Icarus is sometimes the nearest major object to the Sun. At its nearest approach it flies within Mercury's orbit and then hurtles outwards, passing Venus, the Earth and Mars before turning round again near Jupiter. Its orbit is regular and it will be a long time before it comes as near as in 1987. However, close encounters with one or other of the various objects that it passes during its adventures may perturb its orbit such that it will be a hazard one day in the far distant future.

The last time that Icarus approached us, scientists at the Massachusetts Institute of Technology asked themselves what action we could take if some day we discovered that an asteroid was headed directly at us. Hollywood picked up this idea.

The film makers didn't go at it half-heartedly either. They weren't satisfied with a disaster from a single hit. Their story concerned a collision of a comet and an asteroid out in the belt beyond Mars. The collision broke up these bodies into several fragments, some of which headed towards Earth. Scientists first detected the large comet head out in space, and realised that it was coming right at us. The USA and USSR combined to send out all of their nuclear weapons on rockets aimed at the invader with the intention of breaking it up and deflecting it, or at least reducing the effects of its impact.

Building up the tension, the story had some of the small fragments heading towards Earth in advance of the main rock. These small fragments crashed on New York and other major cities (quite remarkable aiming as the odds are that they would land in the sea or in the vast wastes as we remarked in Chapter 2). Finally the central core bore down in the disaster movie to beat them all.

Currently the largest known Apollo object is Hephaistos, which is 6 miles across. In Chapter 5 we will see that an asteroid might have extinguished the dinosaurs 65 million years ago. The suspect was 5–12 miles across. So there is the chance of a life-threatening collision, although the known Apollo objects do not make this very likely; but for one thing – we now have nuclear installations on the planet. If one of those was destroyed the consequences could be very worrying.

In this century we have on several occasions come within days or even hours of hitting an asteroid. The future is not likely to be any different. Every few years one or other of the Earth-crossing Apollo objects will come within hours of hitting us. In time it is certain that we and one of them will score a bull's-eye, but the odds are against it happening in our lifetime.

Meteors

The Apollo objects are outnumbered by the debris of smaller comets which cause the annual meteor showers. We pass through the debris and then voyage around the Sun for a whole year until we come back to the same relative position and pass through the ring again. Although the ring as a whole appears stationary, it is in fact in permanent motion. It consists of pieces of gravel and dust that are themselves orbiting the Sun. One piece of stone rushes past, pursued by another, and another, so that there is a continuous stream of pieces all the time.

Most of the individual pieces of dust in the stream of meteors weigh only a few grams, but the head of a comet might be lurking, several miles across – an Apollo asteroid in the making. Very likely there are many large lumps of rock that could hit and leave a big mark. If so, each year the Earth is crossing a lethal freeway whose traffic travels at 30 miles per second, most of which are gnats but one or two could be boulders the size of mountains.

This is like playing chicken on the roundabout. The Earth is going round and round, missing the big head this time, last time, until eventually the odds run out and we collide.

If a comet has been utterly destroyed then its debris may be distributed uniformly around its entire orbit. We see a meteor display on the same nights each year as we pass through; the intensity is the same from one year to the next. However, if the comet is still around, or only partially decomposed, the density of the debris may vary considerably, being large in the vicinity of the head and negligible far away. In this case the meteor showers will have dramatic peaks in certain years when we pass near to the head while being disappointing in intervening years.

There are a few major meteor streams each year. These include the Leonids (12 November), the Perseids (mid-August), the Beta Taurids (peaking on 29 June) and the Geminids (14 December). Let's look at them in turn.

On average there are about ten meteors per hour when the **Leonid** shower occurs each November. There is a wood-cut in the American Museum of Natural History showing hundreds of trails in a Leonid shower. Even allowing for artistic licence this is excessive, but there must have been a spectacular shower to impress the artist.

It was the great meteor shower of 12 November 1833 that really started serious study of meteors. Hundreds of thousands were seen all over the eastern coast of North America. All of them seemed to radiate from a point in the constellation Leo – hence 'Leonid'. It didn't matter where you were, in New York or the south, the radiant effect was the same point in the sky for all. This showed that the meteors came from outside the atmosphere and their apparent divergence from a single point was simply the effect of perspective: the meteors are really moving in parallel paths.

Records of the Leonid shower have been traced back as far as AD 902. The Norman Conquest was indeed portentious – accompanied by Halley's Comet and followed by a peak Leonid display. The ring of debris has a high-density region that orbits the Sun every 33 years. This is clearly the remnant of the comet head where the final pieces broke up. The nearer we are to the densest bit as we cross the road, the more spectacular the display turns out. In 1833 and 1866 it was very bright but in 1899 and 1932 it was rather poor by comparison, though still brighter than in intervening years. The Leonids were again spectacular in 1966, shooting up to 10,000 per hour at one stage. What does 1999 portend?

In some interpretations of the prophecies of Nostradamus the Apocalypse will occur in that very year. Moreover, a comet will herald it, appearing around 21 June. So the year looks promising, but the date is more akin to the Beta Taurids than Leonids. In addition, the prophecy implies that the comet will come from Ursa Minor rather than Leo, if that gives you comfort.

Whether or not Nostradamus has been correctly interpreted, even assuming he is right, the Leonids are unlikely to be the source of a cometary collision. When tens of thousands of meteors rain in an hour, we are clearly passing through the head of a comet *that has already broken up*. There is unlikely to be any huge core left to strike us.

So much for the Leonids. The **Perseids** in August are more tantalising. These meteors were once known as the 'tears of St Laurence'. They are a steady annual show at the rate of about one a minute at their 12 August peak and extending, less spectacularly, for two weeks on either side of this date. Or at least they were until 1980 and 1981 when the stream suddenly brightened. This may be worrying in 2102 for the following reason – astronomers have lost track of a comet.

The Swift Tuttle comet orbits the sun with a period of 121 years. It was last seen in 1860 and should have returned in 1981 but failed to turn up.

Now a comet can't go off at a tangent any more than the Earth can. It is a part of the solar system and orbits the Sun under the same gravitational forces that hold the Earth and the planets in their courses. If it fails to turn up then it must have died, broken up.

This fits in with the sudden brightening of the Perseids in that very year – the Perseids are its garbage. The big unknown is: is there a big unknown? Is there a huge piece of rock whirling around? In 2102, or 121,242,363 years later, our descendants may find out.

Now you may think that I am trying to impress you with unlikely eventualities. To remove any lingering complacency let's look at the case of the **Beta Taurids** and one of the most devastating natural phenomena this century.

This meteor shower maximises on 29 June. And it was on the morning of 30 June 1908 that a giant fireball streaked across the sky and levelled 7,000 square miles of forest in the Tunguska valley. From reports of the flash and the size of the destruction scientists have estimated that the power of the explosion was like that from a large hydrogen bomb. Various sensational explanations have been proposed: aliens visited in a nuclear-powered spacecraft which crashed; antimatter hit the earth; a mini black hole struck. These all ignore important facts. There is no radioactivity at the site so nuclear explosions by aliens are out. Antimatter would have caused gamma ray bursts and residual radioactive traces. Mini black holes, if they even exist, would have emerged from the other side of the Earth (as they have not eaten up the Earth in the meantime), but there are no records of amazing phenomena in the Atlantic Ocean that day. You can choose the bizarre if you wish to, but there is a natural explanation that fits all the facts.

The consensus among scientists is that a large part of a comet head hit the Earth that day. A comet's nucleus is fragile, being a snowball full of gravel, and as it broke up it left no impact crater even though it caused great damage at the site and disturbed the atmosphere. On this occasion we played chicken once too often and a large piece hit us. This probably implies that the comet associated with the Beta Taurids is now finally broken up and we don't need to worry about it any more.

Finally there are the **Geminids**, which hit us every 13 December. In 1983 a satellite carrying sensitive infra-red detectors (the IRAS – infra-red astronomical satellite) discovered a comet. This was given the codename 1983TB and its orbit was seen to coincide with the Geminid meteor stream.

The satellite discovered it in October 1983 and after watching it for two days plotted its course. It appeared to be heading directly for Earth with impact estimated to be due on 13 December! As more days passed and the comet approached the scientists watched its flight ever more carefully and began to discern that it wasn't exactly in line with the Earth. From far away an object may appear to be heading for you whereas when it is nearer your perspective is better and you see that it will make a near miss. That is how it turned out with 1983TB (obviously, or you would have been aware of it!). It passed between the Earth and the Sun, which is close on an astronomical scale, close enough that maybe you saw it with your own eyes for a few nights.

It wasn't particularly bright and may be nearing the end of its life. Its orbit was calculated and it will periodically return within one-tenth of the distance from the Earth to the Sun, that is within 10 million miles. Jupiter, the most massive planet, will perturb the comet's orbit and pull it away from us. After 2150 it will be totally outside the Earth's orbit and so we have nothing to fear from this one.

However, we can be sure that deep in space there are other comets heading towards us right now. Most of them miss us by 100 million miles or more; occasionally one crosses our orbit, as in 1983.

A 100 metre fragment devastated an unpopulated valley in Siberia in 1908. Where and when will the next one strike?

Nemesis and the Lost Planet

During the 1980s we have learned more about our place in the Universe than ever before. Yet at the same time we have highlighted a surprising area of ignorance about goings-on in our own backyard.

Modern telescopes peer deep into space revealing thousands of galaxies of stars. There are spirals, elliptical shapes, spherical ones and more besides. We can watch galaxies tugging at each other by their mutual gravity, distorting and distending. Within our own galaxy (the Milky Way) we have studied many stars in detail. We know how they are born, how they live and die. Close to home we are rapidly learning much about our solar system, aided by the space missions. Human footprints mark the Moon, robot craft have landed on Mars, we have sent space probes to photograph Jupiter and Uranus, soon to reach Neptune, and have flown through Halley's Comet. There are even plans to send a probe into the Sun. However, we have not yet landed on, or even made a flypast of, an asteroid. So we still have much to learn.

Between the inner solar system and the distant stars there is a real gap in our knowledge: we know almost nothing about the remote outer solar system. The distant planet Pluto was found only 50 years ago. Its sizeable moon, Charon, was unknown until 1978! This shows how little we know of dark objects even when they are fairly near to us.

Astronomers discover about six new comets each year. Occasionally one comes near enough to be visible to the naked eye. The most famous regular visitor is Comet Halley whose pass in 1986 was too remote to be dramatic. In the past it has come much closer; stretching across whole constellations it made an awesome sight. In the future it may come close again.

There must be a huge number of comets in and around the solar system as a whole if a new one is discovered every other month. Jan Oort, a distinguished Dutch astronomer, has estimated that there are 100 billion of them. The orbits of many comets show that they come from a common place deep in space, 50,000 times more remote than the Earth from the Sun, a quarter of the way to the nearest star.

Travel out beyond Neptune and Pluto, further than Voyager has yet reached. The Sun is now a distant dimness. Carry on into the darkness beyond the planets into interstellar space. The next star is still far off. You are travelling through blackness, perpetual nighttime. This is not an empty void; out here in the frozen depths of space comets are icebergs several miles across, invisible from Earth. They have no light of their own – like the rest of the solar system they merely reflect sunlight, and out in this darkness they are as difficult to see as the iceberg that sank the Titanic.

This reservoir of 100 billion icebergs is called the 'Oort cloud'. These cosmic hitchhikers, freeloaders to the solar system, are slowly orbiting the Sun far beyond the planets, so remote that they barely hang on. The slightest disturbance in space could tip millions of these icebergs in towards us.

The existence of craters and other features of the Earth's geological and fossil record suggest to some that we are bombarded by asteroids and comets every 28 million years or so. The last major happening was 11 million years ago, so we are in a quiet period at present. The question is: What is stirring up the distant comets periodically?

The most natural explanation is that the cycle is related to our motion around the galaxy. We orbit the Sun once a year; the Sun and the entire solar system is orbiting the centre of the galaxy every 200 million years (see Figure 4.2). Sometimes we pass near to other

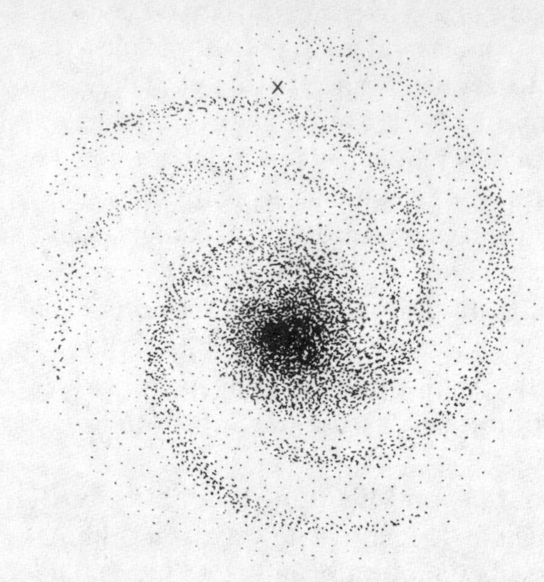

4.2 Milky Way External view of the Milky Way galaxy. The position of the Sun is marked by the X.

stars and they disturb the comets in the outer reaches. Some will be released into deep space, others will fall in towards the Sun.

Two other theories place the blame closer to home. One theory is that in the dark regions beyond the known solar system there may be a tenth planet – called Planet X of course! There is little doubt that something is lurking beyond the outer planets. The discoveries of Neptune and Pluto were stimulated by anomalies in the motion of the planets. Already in the eighteenth century astronomers had noticed that Uranus moved as if under the influence of a distant body. The discovery of Neptune in 1846 explained the perturbation of Uranus but in turn exhibited anomalies of its own. The search began for a more distant massive planet. Pluto was found in 1930 but it is very lightweight. The discovery of its companion Charon in 1978 doesn't really solve the problem; both Pluto and Charon are very small, too tiny to provide the missing tug. Hence the suspicion that a tenth massive planet is still to be found out in deep space. We don't know where or how far away it is with any confidence but we suspect that it is larger than the Earth, some two or three times more remote than Pluto and orbiting the Sun once in 1,000 years. The only clues are that Uranus and Neptune were disturbed in the nineteenth century, whereas in recent decades these planets have been acting 'normally'. This suggests that Planet X was near the planets 100 years ago but is now far out of the plane of their orbits.

Searching for Planet X is a real enterprise. In August 1989 the Voyager 2 space probe will pass by Neptune and head out of the solar system. It will continue to send back signals to scientists at NASA's Deep Space Tracking Network on radio telescopes in California, Australia and Spain. If it, or its sister craft the Pioneers and Voyager 1, should come under the influence of Planet X their trajectories will deviate from those expected. It is very unlikely that in the vastness of space they will chance upon the planet, but it is a possibility none the less.

In 1987 NASA held a press conference at which they reported results from the Pioneer spacecraft, by then in the farthest reaches of the known solar system. Pioneer felt nothing untoward. Pioneer carries instruments that are very sensitive to disturbances. The nineteenth-century disturbances of Uranus and Neptune were big enough to see; the lack of any extra tug on those planets in the twentieth century and the very precise measurements by Pioneer constrain the possible orbits of the tenth planet. Put all this information together and NASA proposes that Planet X is in an orbit tilted relative to those of the known planets (see Figure 4.3). It could be travelling perpendicular to them and could even be the lost moon of Neptune.

Planet X may well exist, but it is unlikely to be responsible for the cometary bombardment. The inner edge of the Oort cloud is at least 20 times more remote and beyond the influence of Planet X in a normal orbit. Encounters with galactic material could disturb the planet's orbit every 28 million years and in turn it would affect the comets. If Planet X exists then it is possible that this sequence occurs. But it seems unnecessarily complicated to invent Planet X for this purpose alone. Why not assert that whatever disturbs Planet X, disturbs the comets directly. A more exciting theory is that the Sun has a small dark companion, a sister star with the romantic name Nemesis (the Greek goddess of doom), moving in a

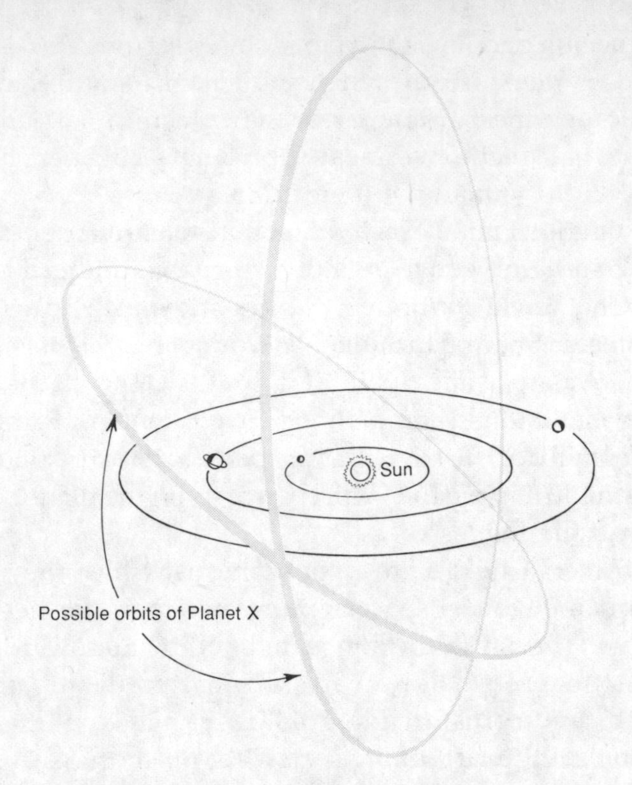

4.3 Planet X – a tenth planet? Possible orbits of Planet X according to NASA in 1987.

long tight ellipse about it. This star is never nearer than 250 times Pluto's distance and reaches as much as 4,500 times as far. With such an orbit it would sweep through the Oort cloud every 30 million years. At present it could be three-quarters of the way to the nearest bright star, Proxima Centauri.

Estimates are that Nemesis is only 1 per cent as massive as the Sun, some 10 times as bulky as Jupiter. This would be a typical example of a dark star, containing too little mass to set the nuclear firers burning. It would give out heat, but not light, like a fire before it glows. So it will not show up in an optical telescope. However, it may already have been detected by IRAS, the 'infra-red astronomical satellite'. If so, information about it is lurking on the data tapes awaiting discovery. It will be 10 years or so before all the data are analysed.

There seems good evidence that something is 'out there' disturbing the comets periodically. As such this is a reminder that even in our own backyard there may be unseen dark objects. There could be a whole plethora of such debris pervading the galaxy and even the entire cosmos, an invisible shadow universe accompanying our own. Such an idea would have been science fiction a few years ago but there are accumulating clues that things may be this way (see Chapter 12). Whether this ghostly matter is responsible for the cometary bombardment is still moot.

When a comet enters the inner regions of the solar system, the gravitational pull of the large planets might destroy it by impact or break-up, might sling it away for ever, or might capture it in a tight path around the Sun.

Some comets are pulled into such tight orbits that they not only cross our path but also graze the Sun. If their orbit is in the same plane as ours they lie in the line of sight of the Sun. This makes it very hard to see them. Only members bright enough to show up in twilit skies are seen. Several examples have been discovered during total eclipses; indeed, enough to make one suspect that there are many more sungrazers yet to be found.

The comet Howard-Koomen-Michels 1979XI was one of these. The cameras on the US Air Force Satellite recorded the comet closing on the Sun at over 150 miles per second. The comet entered just above the plane of the planets, missing us but hitting the Sun. The tail extends further than Venus, over 60 million miles, but on the far side of the Sun and over our heads. After impact the debris scattered over the inner solar system.

So one comet at least has hit the Sun. The Earth is a small target – there's less than 1 in a billion chance that we and a comet will be in the same place at the same moment. But if there are 100 billion comets in the Oort cloud, and if as few as 1 in 100 is disturbed by a passing star, then 1 billion comets have entered the solar system. This makes it almost certain that a comet will hit us some day.

A Halley-sized comet could hit us from the rear or head on. This means that it closes in at anywhere between 10 and 50 miles per second. Taking 30 miles per second as an average one finds that the impact releases as much energy in a second as the Sun brings the entire globe in four months. It is as if all the world's nuclear arsenals were exploded simultaneously in the same place; it corresponds to half a million earthquakes of magnitude 9 on the Richter scale – the largest magnitude ever recorded – all happening at once.

This is enough energy to remove the entire atmosphere of the Earth. This is a dramatic but alarmist way to think of it, since the energy will most probably be dissipated. So we should estimate its effects using other comparisons.

If this energy dissipates as heat throughout the atmosphere it would heat the air up by 190 degrees Celsius. A rise in temperature of the atmosphere would destroy life. High humidity in hot air would condense inside cool bodies as animals breathed it in. Sea animals might fare best if located away from the shock; the oceans overall would warm up by a few degrees only but an enclosed sea like the Mediterranean could boil. The intense heat would vaporise rocks and they would fall back to ground as glassy marbles similar in form to 'tektites' (see Chapter 5). The presence of tektites around the globe may be the smoking gun from catastrophic collisions in prehistory.

A shroud containing billions of tonnes of dust could be thrown in a circle around the Earth, cutting off sunlight. Your bare skin can sense the shadow of a cloud crossing the sun on a summer's day; a dust cloud lasting years could destroy plants and disrupt the food chain. So even though it is alarmist to present it as 'removing the atmosphere', this is hardly more comforting.

This may be what killed the dinosaurs. There has been a lot of detective work trying to identify the culprit. Some scientists now believe that they know the answer.

5

Death of the Dinosaurs

The arrival of mankind and the development of nuclear weapons have created the possibility of self-inflicted Armageddon. It is far more likely that Doomsday will come this way than by a natural catastrophe. With this exception nothing has changed significantly in the natural risk stakes. The chances that we are eliminated tomorrow are pretty much the same as that we were bombarded yesterday. If there is a strong likelihood that we will be struck by a comet or asteroid in the next millions of years, then it is likely that we have been hit in the past.

And we know that we have. The Tunguska event in 1908 and the historic Arizona meteor crater are examples of 'small' impacts. I say 'small' here because a repeat could threaten a city but hardly a country let alone the whole world. What I am more concerned about is the possibility of an entire comet or an asteroid wiping out life as we know it. If that is a real threat in the next 10 million years, say, then major impacts must have happened several times in the past and left their mark in the geological and fossil record.

There are abundant fossil records covering 570 million years, one-quarter of the Earth's existence. During this period there have been five great biological crises when many varieties of organisms disappeared. The most dramatic was the end of the Permian period 250 million years ago, when 96 per cent of all species died out. Towards the end of the Triassic, 215–225 million years past, whole varieties of early amphibians and reptiles disappeared and the dinosaurs first came on the scene in abundance. The most recent of these great extinctions happened 65 million years ago. About one-half of the genera living at that time perished, including marine and flying reptiles, microscopic floating animals and plants, and the most celebrated extinction of all – the death of the dinosaurs. This discontinuity defines the boundary between the Cretaceous and Tertiary epochs.

Within the Tertiary epoch there is sufficient detail in the fossil record that we can recognise subdivisions and it is these that stimulated the first serious suggestions that the Earth suffers bombardment. The death of the dinosaurs is but a recent entrant in the story; it is over 30 years since the initial suggestion that extraterrestrial bombardment has left recent imprints.

Extinctions need have nothing to do with extraterrestrial collisions; alterations in climate, such as occurred in the ice ages, or periods of extreme vulcanism may be sufficient. Indeed the onset of the dinosaurs' dominance is widely thought to be the result of them winning out in a competition with the prevailing species as they adapted better to a changing environment. Rather than a sudden abrupt change, the evidence here points to a relay of competitive replacements of old forms by new ones over a span of 10–25 million years.

Their demise, 65 million years ago, seems rather different however. Very few now believe that mammals competed, stealing the dinosaurs' eggs and thereby eliminating them. Nor did these 100 tonne, 30 metre long beasts become incapable of procreating. Dinosaurs made love delicately and successfully for 150 million years; and participants at the British Association for Advancement of Science annual meeting in 1987 were given a demonstration of how.

Dr Helen Haste bent over a stool and extended her left leg in the air pretending that this was the female dinosaur's giant tail. Beverly Halstead, playing the role of the male dinosaur, then hooked his left leg over hers while keeping his right foot on the floor. The left legs of both were then intertwined. The purpose of the demonstration, apart from being great entertainment and one of the high spots of the entire meeting, was to show how the male dinosaur must keep one leg on the ground to avoid crushing his partner. The juxtaposition was relatively straightforward, and there is no reason to believe that they suddenly became incapable.

Their extinction seems to have been a relatively abrupt chance event and there has been much debate over what caused this.

One of the best publicised is extraterrestrial interference. There are many clues in the geological record that an asteroid or comet the size of Manhattan rammed the Earth, darkening the skies with the dust that it kicked up, blocking out sunlight and killing off plants and animals. This idea, which was originally regarded with some scepticism, leapt to people's attention recently when some physicists concluded that there is evidence that the dinosaurs were killed by the intervention of an extraterrestrial. The attendant publicity has made this explanation something of a 'factoid'. Geologists, palaeontologists, physicists and geochemists now discuss this topic together at conferences. As we shall see, it is still an open question, but as my theme concerns extraterrestrial threats to the planet I shall present their case.

The first clues came from a phenomenon that appears, at first sight, to have nothing at all to do with dinosaurs and extraterrestrials. The story begins with the mystery of the tektites.

The Tektite Mystery

The soil and rocks beneath our feet conceal the history of the Earth. 'The present is the key to the past', wrote the geologist James Hutton in the eighteenth century, and as more and more geologists studied rocks in the field it became clear that the Earth's surface took millions of years to form. Volcanic eruptions, the advance and retreat of the seas as the land rose and fell, the uplifting and eroding of mountains, left layers of rocks that are today far underground but were once at the surface. In mountainous regions, in cliffs and gorges, the ages are revealed. The fossil remains in these rocks tell us about the plant and animal life of past epochs.

The Earth is made of rocks, which are in turn made of minerals – inorganic chemicals (inorganic meaning that it is not like an animal or plant). Familiar examples include gypsum, quartz, topaz and diamonds. Rocks are classified according to their origin. Igneous rocks are the result of molten lava that has cooled. Sedimentary rocks result from sediments, such as sand and pebbles, being compressed into solid bulk such as sandstone, clay and chalk. Metamorphic rocks are rocks that have been changed in some way from their original state, for example being worn down like soil.

Many rocks and minerals have been fashioned into tools, such as stone, flint and iron in early societies, or treasured for their beauty, such as gold and diamond. Among these have been curiosities, such as tektites.

Tektites are dark natural glass marbles (the word 'tektite' comes from the Greek 'tektos' meaning melted). The first examples turned up in Moldavia, Czechoslovakia, in the eighteenth century. They were a deep bottle-green colour and could well have been from a long-forgotten bottle works except that they covered too wide an area. 10,000 pieces, most weighing 10–20 grams, were strewn around.

Since about 1860 several sites have been discovered around the world, with the same lack of explanation. They are named according to their location, thus Moldavites in Bohemia, Darwin glass in Tasmania and Libyan Desert Glass for example. They arrived at widely different times: for example, North American tektites are some 35 million years old; European tektites are 14 million years old; while the Australian and Ivory Coast tektites are more recent, between 700,000 and 1 million years old.

The tektites are all natural glass, weighing as little as 1 gram or as much as 8 kilograms. Most are black, some are green, a few are yellow. They are lustrous with a delicate sheen, and have ridges and furrows which follow a contorted inner structure. They come in many shapes – spheres, dumbbells and discs. Their allure has caused them to be cherished as charms in some parts of the world.

Their chemical make-up is similar to sedimentary rocks, but no known Earthly process can account for the transformation of those sediments into glassy balls. At first people thought that tektites came from volcanic eruptions. However, they turned out to have a different chemical composition than the familiar glassy materials from such eruptions. Moreover, there were no obvious volcanic sites near to the tektite fields. Most of them are small and round, suggesting that they fell through the atmosphere, but no one has ever seen a tektite fall, nor are they found in meteorites.

Suggestions for their origin have included lightning in a dusty atmosphere, the heating of stony meteorites, the collision of a meteorite with a natural Earth satellite (unspecified as to what this was), and the fall of antimatter. The operative word is 'fall'; these ideas fell if ever they were regarded very seriously.

Tektites have always been a major puzzle to people who 'pick up rocks and think'. By 1982 there were two major contenders, both involving extraterrestrials. One was that they are volcanic products from the Moon, the other that they are due to the impact of comets or asteroids hitting the Earth.

The Apollo Moon craft brought back lunar samples. As a result of the analysis of the lunar rocks it seems increasingly unlikely that lunar volcanoes are the explanation of Earthly tektites. Supporters of lunar origin theory are in retreat and the impact theory is gaining ground.

Harold Urey's Theory of Cometary Bombardment

In 1939 L. J. Spencer had suggested that tektites are the result of meteorite falls. As there were no obvious craters associated with them it was supposed that the meteorites broke up before impact.

Years later Harold Urey (best known for his 1934 Nobel Prize for the discovery of deuterium – 'heavy hydrogen') started to think about the problem from a physicist's perspective and decided that tektites don't come from outer space. First, he realised that they could not be due to the break-up of a large mass. Such an occurrence could rain down marbles over a few miles but couldn't explain the presence of tektites all over southern Australia. Nor could they have arrived in a ready-made diffuse swarm; the gravity of the Sun would have broken up the swarm and spread tektites all over the Earth. The meteorite break-up didn't give us enough; this other extreme gives too much.

A dense swarm could avoid the global spreading at the expense of giving much too dense a distribution of tektites on the ground. Tektites are spread diffusely over a large area; they do not cover the globe uniformly nor are they piled up to 100 grams per square centimetre. Urey was foxed. As a 'last resort' he thought that there might have been a collision between the Earth and a minor planet (asteroid) and that some unknown mechanism had covered over the crater, as the tektite sites seemed to imply landfall.

This isn't a very satisfactory explanation. Urey remained puzzled. Then the approach of the comet Arend-Roland set him thinking: what would happen if a comet hit the Earth? Apparently no one had worked out the consequences of this in any detail, so Urey set about the task. Suddenly everything began to fall into place.

Urey made his calculations years before the Giotto mission to Comet Halley showed us what the inside of a comet is really like. Giotto's results imply that Urey could well be right.

The comet's head is a loose structure that is being buffeted by high-speed particles coming from the Sun (the solar wind). The effect is as if the material in the comet's head is slow-burning high-explosive chemicals. Urey calculated how much explosive power there is inside the comet. If its head is 7 miles across, and it is moving at 25 miles per second (typical velocity of impact meteors or any body moving under the influence of the Sun's gravity at Earth orbit), then it contains the energy of 50 million atomic bombs. This energy is released when it hits the atmosphere. At such a speed the impact between the comet and the atmosphere is like hitting a solid barrier. Compression and heating make the comet explode. Most of its mass is disrupted into a high-speed hot gas which continues to Earth, heating the surface to melting point.

It is easy to describe such extremes as 'like so many hydrogen bombs', but this isn't quite accurate. The disruption of a comet begins 40–60 miles up and has some similarities to a high airburst of a nuclear bomb. However, it has rather different effects because of the compressed air and relatively low temperatures distributed through its mass. A nuclear explosion is compact, millions of degrees and emits lethal radiation. Disruption of a comet is much cooler than this, with temperatures similar to the surface of the Sun (still hot in our parlance but trifling on the nuclear scale). The effects are more like a propellant than a detonating type of explosion.

The mass hits the surface and is stopped within a second. The maximum pressure can be 40,000 times greater than atmospheric. A broad area is involved, with the result that no deep penetration results. The qualitative features of the Tunguska event are akin to a feeble version of this.

The surface temperature at the impact is a world where rocks vaporise. The blast ejects this maelstrom up into the atmosphere where it cools into glassy pellets. It is these little marbles that we find on the ground and call tektites. So tektites are like the smoking gun. They come from the Earth's surface: rock that melted and cooled again. The theory explains why tektites have a varying composition – they result from the local rocks. The impact provides a natural high temperature for melting and the markings on many tektites are like those expected from cooling solidifying liquids. The occurrence of tektites over a wide area is also explained.

Not only does this explanation fit with the qualitative features but it also agrees well with the quantities of tektites and the expectations based on our knowledge of comets. This is important. Many pseudo-science 'explanations' fail when one confronts them *quantitatively*; apparently nice ideas sometimes just do not fit the scale of the phenomenon.

If the tektites are really strewn about as the result of an impact, then we can deduce the size of the invader from the extent and amount of tektites it spawned. The North American tektites are found in an area covering millions of square miles. The sums imply that the invader must have weighed at least 50 billion tonnes, its girth being several miles. This is an underestimate because the size of the fallout region could be greater; there could be tektites undiscovered in the oceans for example. This mass is quite reasonable for a comet; Comet Halley, for example, is some 20 times this size.

If the tektites are caused by collisions, then their ages suggest that four major collisions have occurred over some 35 million years. This fits with random chance and is easy to calculate.

Many comets come within the Earth's orbit each year. We have to estimate the chance that one of them, and us, are in the same place at the same time. Comets come from all directions, below or above the plane of our orbit, so the chance of collision is much less

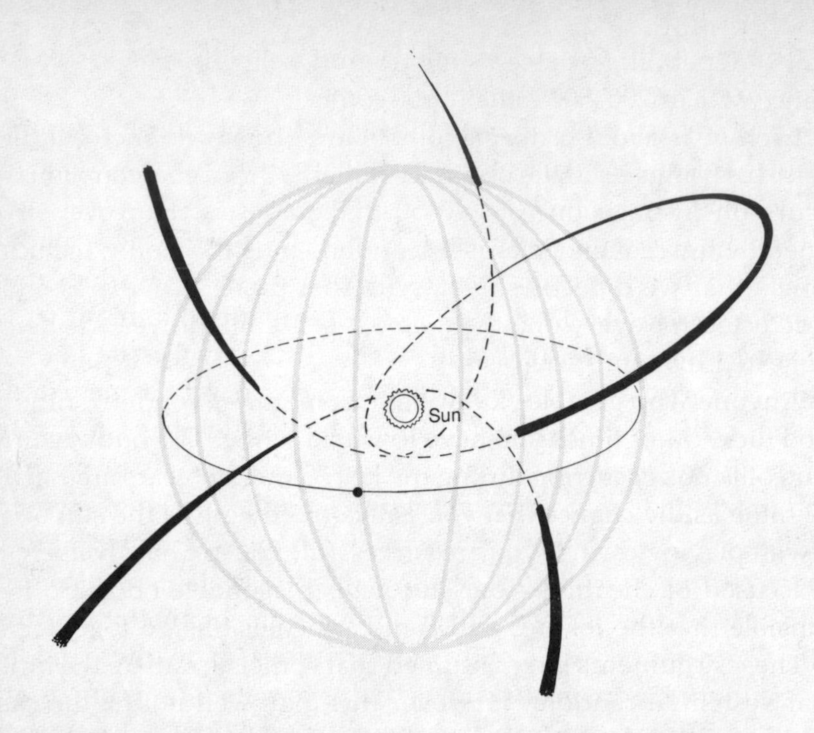

5.1 Earth – crossing trajectories The Earth is represented by the black dot travelling around the Sun on the near-circular path, radius about 100 million miles. I have shown a sphere of this radius. Any comet or other heavenly body that comes from remote space to approach closer to the Sun than we do must cross this sphere.

than if they and we were travelling on the same flat plate. Instead of thinking of the circle that the Earth orbits around, we should imagine a sphere whose radius is the same as the Earth's orbit. We are interested in comets that come from without and enter the sphere.

At the moment a comet crosses the sphere, the Earth is somewhere on the surface. Imagine it as the little black disc in Figure 5.1. The chance that the comet hits the Earth is in the ratio of the area of the disc to the area of the sphere's surface, namely 1:2,000,000,000. In fact, the chance is twice as big as this because if the comet misses on the way in it still has a chance to hit us on the way back. The earth's gravity reaches out into space and can ensnare comets that might otherwise have missed. This is not a dramatic effect; the Earth isn't like Jupiter for instance, but it does increase the odds by

2 to 5 (depending whose estimates you believe). The result is a chance of 1 in 200–500 million per comet.

Each year about a dozen comets are known to encroach like this. This brings us to the original estimates that a cometary impact occurs on average once in 20–50 million years. However, it is generally agreed that this is an underestimate. It only includes comets that we detect; estimates are that there as many as eight undetected for every one that is. So 1 in 10 million years is being suggested by some scientists.

Anyone who gambles knows that sometimes you have runs of good luck, other times you have to wait a while. The odds against four collisions occurring during the last 35 million years are 1 in 4 – the same as the chance that you will correctly guess the suit of an unseen playing card. If you are successful, you are hardly likely to be accused of cheating – one in four is reasonable odds. So it is plausible that the tektites and cometary collisions go together.

These arguments have assumed that comets irradiate the inner solar system at random. However this may well not be the case since we are journeying around not just the Sun but also the whole galaxy. Conditions change and the chance of tipping comets is bigger at some periods than others.

The events that we have been concerned with are on the scale of global catastrophes. They involve collisions with pristine comets. Comets break up as they journey repeatedly around the Sun – Halley's comet is getting smaller each time as material trails off in its tail. A comet that has broken up has a much greater extent and greater chance of hitting us, but the consequences are correspondingly less severe. The Tunguska event is an example. At the extreme we meet the debris from dead comets every year in the annual meteor showers.

So when we ask 'What is the chance of hitting a comet', we are trading off the size versus the chance. We hit small pieces every year. Whole comets every 10 million years or so. Once in 100 million years we may hit a real monster. Then the consequences could be truly catastrophic. Urey's calculations imply that the dissipation of the impact energy could vaporise the oceans and seriously affect the climatic conditions across the entire globe.

EPOCH		TEKTITE	
Pleistocene			
	1	Australites	0.7 ± 0.1
		Ivory Coast	1.2 ± 0.2
Pliocene			
	13	Moldavites	14.7 ± 0.7
Miocene			
	25	Libyan Desert Glasses	28.6 ± 2
Oligocene			
	36	Bediasites	34.7 ± 2
Eocene			
	58		
Palaeocene			
	63		
Cretaceous			

5.2 Ages of tektites compared with epochs (as of 1973)

Comets and Recent Geological Changes

The origin of tektites fits well with Urey's theory of cometary invasion. But is it correct? The impact of a comet can devastate rocks, so explaining tektites, but will disrupt a lot more besides. Surely the geological record must show evidence for these singular occurrences.

Urey first published his idea in *The Saturday Review of Literature* but later commented that 'no scientist but me, as far as I know, reads that magazine'. As a result he developed the idea in the leading scientific journal *Nature* in March 1973.

Urey proposed that tektites are just one result of cometary collisions. He went further and suggested that these impacts caused the major changes that we now identify as the end of the geological epochs. As far as I know this was the first *quantitatively* evaluated proposal of such dramas. Alvarez's later studies of the Cretaceous–Tertiary boundary and its connection with the demise of the dinosaurs stem in part from this. Indeed, Urey explicitly suggested such a connection with a cometary collision as the source of the dinosaurs' demise, based on the (apparent) success of his confrontation between his theory of tektites and the end of geological epochs. (I say 'apparent' because developments since the appearance of his theory call this part of it into question.)

In the comparison in Figure 5.2 the geological periods with the transition times in millions of years are listed in the left-hand column and the ages of the tektites are listed on the right. These are the data as they appeared in Urey's paper – the tektite dates originate with a 1971 paper by S. Durrani and the geological ages came from the standard 1961 work of J. Laurence Kulp.

There is indeed a tantalising correspondence between tektites and the ending of recent epochs. The Australites and Ivory Coast samples coincide with the most recent change. The Moldavites and Libyan Glasses almost coincide with the Miocene to Pliocene and the Oligocene to Miocene transitions respectively. The Bediasites are right on the end of the Eocene period.

That is how things stood in 1973. However, since then there have been some important revisions in the accepted dates of the geological epochs. The ages of rocks, and tektites, are fairly well agreed upon and have changed only very slightly if at all. However, as knowledge of rocks has improved so has the ability to subdivide them, and the agreed boundaries between epochs have changed in some cases. The boundary between Pleistocene and Pliocene has moved back to 1.6 million years, making the correlation with the Australite and Ivory Coast tektites less impressive. The boundary between Miocene and Pliocene has changed dramatically, moving up from 13 million to only 5½ million. The Moldavite tektites, at 15 million years, bear no correlation at all with the change in epochs. (See Figure 5.3.)

It is still possible to make a case for tektites being the smoking guns of collisions with extraterrestrials. However, it is quite another matter to go further and correlate them with other major geological events. The most celebrated attempt to do so involves the Cretaceous–Tertiary epochs and the death of the dinosaurs.

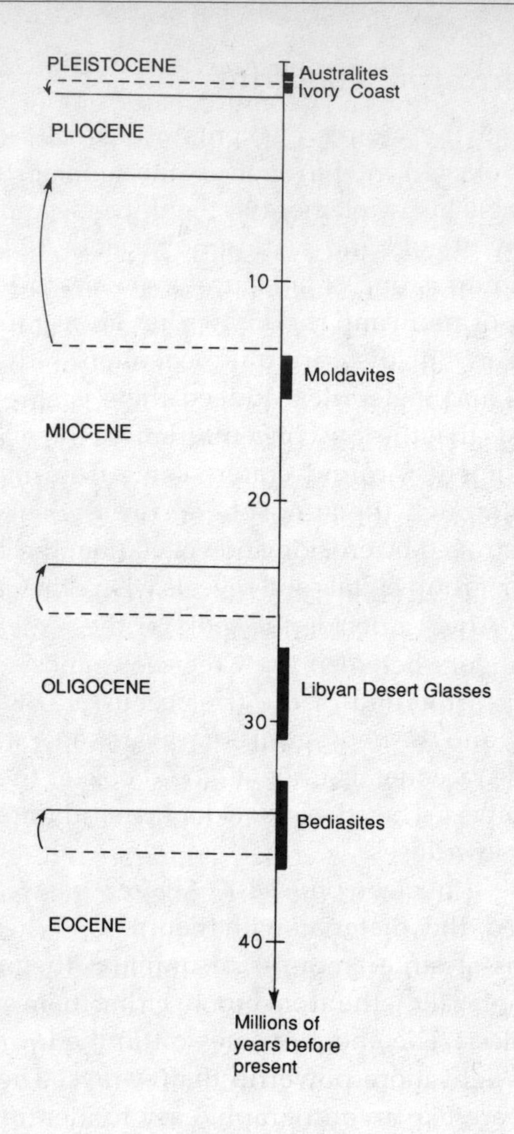

5.3 Comparison of tektites and epochs (as of 1987) Major tektite fields are shown on the right according to their ages (in millions of years) including uncertainties in those ages. On the left they are compared with the transitions between epochs. The dashed lines show transition dates from Kulp (1961); the solid lines are from Snelling (1987) and the arrows show the changes. The correlation noted by Urey between Kulp's transition dates (dashed lines) and the ages of the tektites (in solid black) has vanished.

Extraterrestrial Footprints

Not all of the atomic elements are equally common on Earth. Carbon, oxygen and nitrogen are the familiar stuff of life; iron is the most stable heavy element in the universe and is the stuff of rust; the rarity of gold and platinum makes them valuable.

Platinum is one of a group of similar rare metals – platinum, iridium, osmium and rhodium. They are as rare as their names are unfamiliar. But they are quite common in average solar system material and meteorites. Indeed there is agreement among geophysicists that the low concentrations of the platinum group in soil and sediments around craters come from meteorites that have passed through the atmosphere. Even where craters have been long destroyed by erosion and vegetation, the concentration of the platinum group metals still reveals where meteorites have landed.

Luiz Alvarez decided to look for these metals in sediments at the boundary between the Cretaceous and Tertiary layers. If the rocks from the end of the Cretaceous era, where the dinosaurs roamed, and the new age (Tertiary) are separated by clays containing these rare metals, then it would bolster the prosecution's case for an impact. They decided to look for iridium as it is easy to detect even in low doses.

Being a nuclear physicist, Alvarez knew how to do this; he irradiated the material with neutrons. When you direct slow neutrons at large amounts of uranium they can release energy catastrophically – the atom bomb. Firing them at iridium, however, is harmless. The energy comes off in the form of gamma rays, a form of light more powerful than X-rays. The gamma rays from iridium are like an autograph, easy to identify.

This is such a sensitive technique that you can detect iridium even when it is present in only the minutest traces. You have to be very careful to ensure that there is no other iridium around that could contaminate the delicate search. As iridium is so rare you might think that this is a needless worry, but this is not so.

In 1981 Alvarez thought that he had evidence for iridium deposits in samples from the Cretaceous–Tertiary boundary in Montana. In fact the iridium didn't come from this at all but was in a wedding ring worn by one of the technicians who prepared samples for analysis. Platinum for jewellery contains about 10 per cent iridium as a hardening agent. If a platinum ring loses one-tenth of its mass in 30 years, which is fairly typical, then the average loss each minute is 100 times greater than the measurement sensitivity of Alvarez's experiment. Tests showed that the same is true for gold rings. It is very easy to contaminate the samples and Alvarez has taken great care about this. There have been disputes as to whether there is an effect or not, but it is now agreed that there almost certainly is. This is how Alvarez found the evidence.

In the Apennine mountains of northern Italy there are exposures of sedimentary rocks representing the period from Early Jurassic to Oligocene, from 185 to 25 million years ago. These contain the end of the Cretaceous and beginning of the Tertiary epochs in the form of pink limestone interleaved with clay. The abrupt change in the nature of life is clearly visible when samples from the boundary layer are put under a microscope. In the Cretaceous layer for instance there are foraminifers that are about 1 millimetre in size. These disappear abruptly across the boundary and are replaced by a different genus, of less than one-tenth of a millimetre.

In well-exposed sections the limestone beds from the two epochs are separated by a band of clay about 1 cm thick. There are no fossils in the clay so you cannot 'watch' the extinction occur. Instead it is a case of 'now you see it, now you don't'.

Alvarez took samples from the boundary layer in the Bottaccione Gorge near Gubbio in Umbria, Northern Italy. The gorge reveals rocks through a span of nearly 400 metres covering the passage of the ages. He chose various samples from the Cretaceous rocks, some from the Tertiary rocks and the crucial samples from the boundary layer. The main question was whether or not the chemicals in the boundary layer were markedly different from those in the other rocks, and in particular whether the tell-tale signs of iridium were there.

He concentrated on 28 elements that showed up clearly in the samples. The relative amounts of 27 of them were very similar throughout the whole range of rocks, representing 155 million years of time. It was the 28th that stood out anomalously: this was iridium. There were only faint traces in the rocks below and above the divide, but as you crossed the boundary the iridium count shot up by a staggering factor of 30. None of the other elements as much as doubled in the boundary layer. There was no doubt – a lot of iridium was deposited at the time of the disaster.

This is as expected if the extinction was caused by extraterrestrial invasion. However, you can't immediately infer that here you have absolute proof. There was the possibility that fate had played a cruel trick and that some anomalous conditions in the vicinity had caused the iridium to deposit in those crucial centimetres of the gorge that happened to coincide with the boundary layer.

Fortunately the Bottaccione Gorge is not the only place where the Cretaceous–Tertiary boundary layer shows through. To test whether the iridium anomaly was just a fluke, Alvarez analysed sediments of a similar age from other sites.

One of the best-known places where the boundary layer shows up in Northern Europe is the sea-cliff of Stevns Klint, some 30 miles south of Copenhagen. Here the Cretaceous layer consists of white chalk and the boundary layer is marked by a 35 cm thick deposit of clay. At the Danish site they were able to study 48 different elements, more than half the number that occur in totality on Earth. As in the Italian sample, the iridium increased dramatically in the boundary layer. Indeed, in the Danish sample the iridium increased more than a hundredfold.

Similar results have been found in New Zealand, where iridium in the boundary layer at Woodside Creek is 20 times as intense as in the remaining rocks. There is no doubt that the effect is real and worldwide.

Some geologists argue that certain volcanic eruptions spew out magma whose elemental abundances are like this. (There is a good and detailed survey of this in the 1987 *Nature* article by C. B. Officer and co-workers.) Part of the debate centres on whether the iridium was laid down over an extended period, as in the volcanic theory, or suddenly, as in an impact.

One of the features of the boundary layers in both the Italian and Danish sites is the existence of a layer of clay about 1 cm thick. Alvarez has suggested that it is made of material that fell out of the stratosphere following the disturbance of the impact. Geologists counter by saying that this is naïve; geology is no less complicated a science than physics – the physicists have their physics right but that doesn't make them experts in geology. It is possible that the clays have quite 'ordinary' origins.

So we are left with two questions, each of which is still being hotly debated: were the dinosaurs killed by this catastrophe; and was the catastrophe caused by an extraterrestrial blast?

If one concentrates on the demise of the dinosaurs to the exclusion of all others then you might happily believe that the broad features of the extinction fit in with the idea that there was extended darkness at noon owing to dust in the atmosphere, whether from impact or eruptions. However, the globe contained much more than dinosaurs at that time and this gives many worries about the details. For example, tropical plants managed to survive though you might have expected that they would suffer from lack of light more than most. One palaeontologist tells me that he has fossil reptiles weighing several tens of kilos that survived the catastrophe. How? Why were the extinctions so selective?

Critics of the impact theory point out that there are hints that the extinctions weren't sudden but were spread over some thousands of years – though at a remote 60 million years a few thousand might simply be due to sampling errors.

However, if there was not an impact, Nature certainly conspired to make it look as if there was one. First, if the iridium in the boundary layer came from an asteroid, then you can estimate its weight. It comes out at 1 billion tonnes or, in a more visualisable form, the lump of rock would have been some 5 miles in diameter. Second, if the clay is the result of the dust thrown up by the impact, you get a similar estimate of the rock's size that caused this mess. Third, other noble metals in the Danish clay, such as gold, nickel and cobalt, appear far in excess of their usual terrestrial abundances, suggesting a rock size as much as 8 miles across. All these figures come out the same; and there is no doubt that if we were hit by a blast like that it would certainly disrupt the environment in a catastrophic manner.

The evidence may be only circumstantial, but after 60 million years it is something to have any evidence at all. I know a respected physicist who believes that the case is established, and a well-known palaeontologist who asserts that the verdict is not guilty. My own verdict is 'not proven'. One thing that we would all agree on is that research into this question should continue; it fascinates and stimulates youngsters to become scientists and may be a sobering reminder of what can happen to us when the Earth's environment is drastically altered, be it from extraterrestrial causes or self-inflicted. The dinosaurs died out after 150 million years, the dominant creatures of their time; we have lorded over the Earth for a mere 1 million years – there is nothing to guarantee that we are here for ever.

PART 2

The Nearest Star

6

The Sunshine of Our Life

We live six miles beneath an umbrella of air, water vapour and assorted gases that we call the atmosphere. Climatic changes occur as the Earth's energy balance is disturbed. The Sun is the main energy source, driving the atmospheric circulation and controlling climate. Sunlight hitting the upper atmosphere is scattered, absorbed and re-emitted by molecules and undergoes many changes en route to the ground and to the circulating low-level air that immediately affects us.

The Earth's surface plays a role. Seas act as heat reservoirs; mountains and valleys affect the air flows. The whole interaction is very complex and meteorology has only recently become more reliable with the advent of remote sensing satellites. Any disturbance can trigger significant climatic change. Automobile exhausts and aerosols are changing the atmosphere's chemistry and altering the heat balance. If the Sun's output changes or its radiation is intercepted, the effects on the Earth's climate may be dramatic, as in the ice ages. As the Sun is the main supplier of energy I will concentrate on its behaviour and how it can affect us.

A year sees us journey once around the Sun and experience the range of variety on offer. The surface of the planet spins 1,000 miles each hour and in 24 hours we rotate once, in and out of the Sun's shadow and blaze. The atmosphere warms and cools, perturbs and settles down again.

Day after day we orbit the Sun. Our path isn't a perfect circle but is distorted slightly into an ellipse. We are slightly nearer to the Sun in January than in July, so the net heating at the turn of the year is more than in the middle. This is only a minor effect relative to that from the tilting of the poles, which present us towards the Sun placing it high in the sky during the summer and then tip us away, giving a low sun during winter months.

By the time that we have tilted up and down once, we have circled the Sun once. A year has passed. Some points around the orbit are special because we pass through fellow solar travellers – the rings of rocks that burn up in the atmosphere as meteors. And that is how things would be year after year if everything else was equal. There would be statistical fluctuations – extra-hot days, extra-violent storms – but nothing singularly new.

But that is *not* the whole story. We are orbiting a star that is itself making a grand tour of the galaxy. The Sun takes 200 million of our years to complete a galactic year and it takes us along with it of course. During this circuit we pass through a variety of scenery. There are quiet periods, as now, when nothing much is around; there are times when other stars get rather near; and periodically we pass through the galaxy's spiral arms – dense clouds of dust where new stars are being formed. Whereas the Earth passes through the solar rings of rocks in a few days, the galactic encounters are scaled up. It takes 1 million years to pass through the dust clouds and they can dim the light, get into the Sun, modifying its heat generation, and change the Earth's climate markedly in the process. Some astronomers, but by no means all of them, have argued that the great ice ages are a result of this: the galactic analogues of our annual meteor showers.

Our next encounter with a spiral arm is far off and as far as we can tell the road we're on is fairly clear for the next few billion miles. (We're passing through a fine mist of intergalactic dust at the moment but it isn't noticeably affecting us.) However, there is the possibility that changes within the Sun might affect us in the short term.

We take the Sun for granted yet anything that cuts it off from us will give us traumas. The prophets of nuclear winter have spelled this out clearly; we may be able to hide away from the nuclear radiation by burrowing underground like moles, but the burning forest and cities will blanket the globe in smoke and cut off sunlight. Many scientists believe that this could be the end of the human race. Even if we don't inflict darkness on ourselves, Nature could switch out the light.

I remember as a child seeing a spot on the sun one misty day and being astounded at my discovery. All the children in the playground were looking at it and rushed out from their classrooms at lunchtime to see if it was still there. It was, and in my imagination seemed bigger than before. I was certain that this dark spot was growing and would blot out the Sun. The thought that the Sun might disappear instilled panic of the kind that primitive societies must feel when they see the Sun being eaten up at total eclipse. Like them, I needn't have worried. Sunspots appear regularly, peaking every 11 years or so.

Sunspots have been blamed for many things. There are claims that they affect the weather and even the results of political elections. Changes in the Sun can leave a mark on the Earth – literally. These marks enable us to see how the Sun has behaved over 700 million years.

The Sun is the nearest star to Earth. It is close enough that we can watch it in detail and learn how stars work. Our whole existence depends on it, so the better we understand its workings the better placed we are to contemplate our immediate future. The problem is that our best theories don't fit everything perfectly; there are some disturbing hints that the Sun, or something, is misbehaving. How much do we know?

Measuring the Sun

We know how far away the Sun is. In Chapter 3 we learned how that measurement is made; it involved bouncing radar from Venus to see how far away that planet is and then comparing the time that Venus takes to orbit the Sun as compared to our own year. Put those two pieces of information together and the distance to the Sun comes out to be nearly 100 million miles.

The Sun is about as big as a thumbnail viewed at arm's length. The Sun is 100 billion times further away than that, so we can 'measure' its diameter as 100 billion thumbnails – that is about 900,000 miles, or about 120 times the diameter of the Earth.

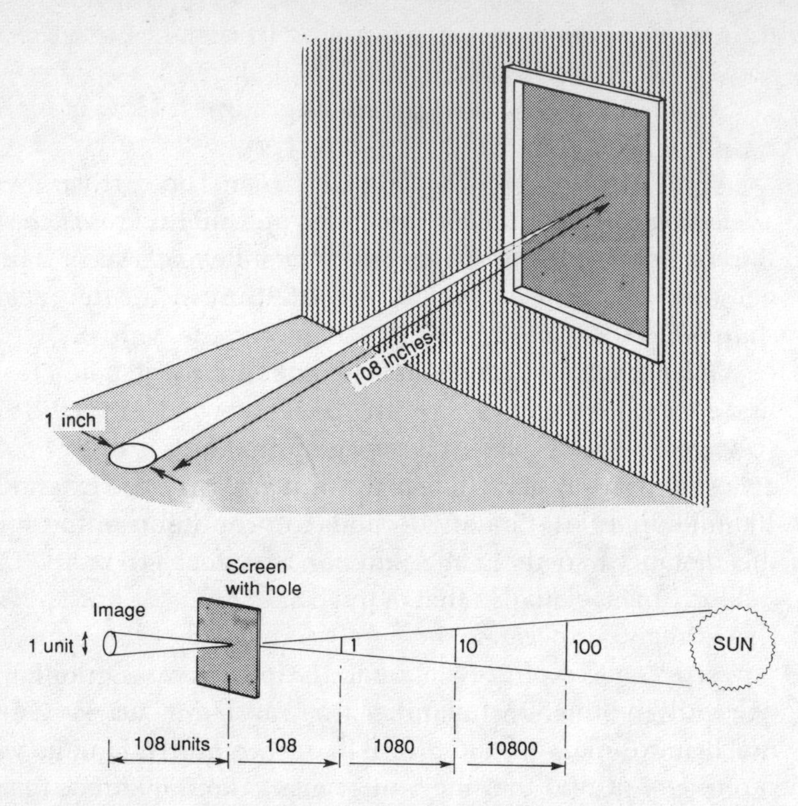

6.1 The size of the Sun (a) Measuring the size of the Sun through a pinhole in a blind.

(b) The Sun's diameter is 1/108 of its distance from the Earth. This scale of 1 to 108 is seen in its pinhole image whose size is always 1/108 of its distance from the pinhole. Simple geometry shows this is true of the actual Sun relative to its distance from us.

A simple experiment will give you a more accurate measure of the Sun's size (see Figure 6.1). On a sunny day cover a south-facing window with a blind and make a pinhole in it. On the floor you will see an image of the Sun. Measure its size accurately and also measure the distance to the pinhole. You will find that this distance is 108 times greater than the diameter of the image. It doesn't matter how big your room is: in a small room the Sun casts a small image while in a big room the image is correspondingly bigger. If your room was nearly 100 million miles big, the size of the image would be the same as the real size of the Sun. Simple geometry shows that the image and the distance will always scale like this. So the Sun's diameter is 1/108 as big as its distance from us, hence some 900,000 miles (so the thumbnail sketch was pretty good).

So we know how much space it occupies. How much does it weigh?

Apples and hailstones fall to the Earth. The Earth and we on it are falling into the Sun at a rate of about 3 millimetres each second. We are also moving forwards at several miles each second, with the net result that we travel around the Sun in a huge near-circle completed in a year.

When an apple drops from a high branch it will fall 16 feet in the first second, which is 1,500 times as far as we fall into the Sun in a second. This is because the centre of the Earth is only 4,000 miles away from the apple whereas the Sun's gravity has extended over 100 million miles. Gravity declines in proportion to the square of the distance, so the Sun's gravity is enfeebled more than 500 million times relative to that of the Earth. That it is only 1,500 times less in reality is because the Sun is so massive – it is a much more powerful source of gravitational attraction than is the Earth. Indeed, divide the 500 million by 1,500 and you can work out how much more massive the Sun is than the Earth. Doing the sums carefully you find that the Sun's mass is 330,000 times that of the Earth.

The volume of the Sun is about 1 million times that of the Earth and so we deduce at once that its mean density is about one-third of ours, or two to three times that of water. But don't get the impression that the Sun is like this right through. It is a ball of hydrogen gas which is very tenuous on the periphery and conversely very dense indeed at the centre. To maintain such a situation the centre of the Sun must be incredibly hot, over 10 million degrees. We can infer that the temperature of the outer surface is only 6,000 degrees. So already we have deduced quite a bit about the Sun from a range of 100 million miles.

Our current theories give the following picture of the Sun's inside. A quarter of its radius contains an active core in which thermonuclear reactions produce solar energy: protons (the nuclei of hydrogen atoms) fuse together to build up heavier elements and produce ghostly particles called neutrinos as byproducts, which stream out into space (see Chapter 7). Above this core (up to 70 per cent of the Sun's radius) extends a zone where heat radiates upwards. The outer 30 per cent is called the convective zone where the heat is transported by turbulent motion of the gases.

In this heat atoms are unable to hold themselves together and are disrupted into their component parts – negatively charged electrons and positively charged nuclei. The swirling motion of these electrically charged particles creates intense magnetic fields whose structure alternates on a 22-year cycle. It is like north magnetic pole flipped to south and back again, with the result that if you measure only the strength of the field but not its direction you will perceive two 11-year cycles of intensity. The strength of the magnetic fields can vary tenfold during a cycle, being least in years when the Sun appears quiet; in active years the magnetic field breaks through the surface and gives rise to sunspots and solar flares.

Sunspots

There is a deep-seated psychological belief that the Sun is changeless and perfect. The Sun's regular rising and setting have been watched, even worshipped, for millennia. The remnants of that worship are still with us today. Druids visit Stonehenge, almost certainly an ancient computer for measuring the seasons – survival depended upon sowing and reaping at the right time. Even the date of Christmas Day, 25 December, is a reminder of an ancient festival, namely the rebirth of the Sun following midwinter.

The Aztecs worshipped the Sun. Visit the beaches of the Mediterranean or southern California and you will see modern adherents to the cult, lying full length beneath its rays.

The circle of gold in the blue sky symbolised perfection, God's creation – the Church persecuted Galileo when he first reported seeing spots on the Sun. The desire for perfection and constancy may have led us to see such where it is not warranted. Nor is the Sun constant – it wobbles and currently is fading at a rate that could produce a mini ice age in another 50 years. This is not as bizarre as you might think; the Sun's variability has caused climatic changes lasting decades in recent times.

We may complain about the weather but however bad the winters they are nothing like 300 years ago. A mini ice age hit northern Europe. Glaciers that usually end high in the mountains encroached on villages in the alpine valleys. The River Thames froze and there are many paintings from the seventeenth-century Dutch school showing ice skaters on the dykes of Holland. During this period something unusual was taking place on the Sun's surface, noticed by Europeans, Oriental and mid-Eastern astrologers.

A meeting of the US Geophysical Union early in 1986 brought together a number of independent scientific studies which all agreed that the Sun has steadily declined since 1979. Satellites in earth orbit, rockets, balloons and measurements at ground level all showed that the Sun is fading. This appears to be a continuing trend on top of which is a short-period cycle which may be related to the sunspot cycle, but there is no definitive knowledge yet whether these are related or not.

Sunspots come and go, small spots lasting for only a few hours while some large ones as big as the Earth can survive for months. In peak years as many as 200 may occur. In between the activity dies and only half a dozen spots might occur in the minimum of the cycle.

The amount of heat output – the Sun's brilliance – drops in proportion to the amount of the surface that has been masked by the spots. In terms of the *total* radiation arriving at the Earth, this is a small effect and the balance is restored later. However, there is still a lot that we don't know about the dynamics behind this. Is there more X-ray and less infra-red, or more ultra-violet and less radio emission? It is as if a great orchestra dimmed a chord – did all the strings diminish or did the double basses quit while the violins increased a little? We need to know more about this because the different parts of the electromagnetic spectrum affect the chemicals in our upper atmosphere in different ways.

There has been a lot of debate as to whether or not sunspots correlate with our weather. If you want to pursue that debate read the articles cited in the Suggestions for Further Reading. These correlations are often controversial though the prolonged cold snap in northern Europe coincided with the 'Maunder Minimum' between 1640 and 1720 (named after E. W. Maunder, superintendent of the solar department of the Greenwich Observatory). During that period the Sun seemed spotless. Some people claim that no one was looking, but that seems facile; the astronomers of the day were as good as any and the minimum of solar activity seems real.

It may seem fanciful that distant spots can have any effect on us here. However, although the Sun's shining face is very remote, its atmosphere extends far beyond the Earth; we are orbiting *inside* the Sun!

Magnetic fields within the Sun are being twisted in knotty contortions. They suck in the bright surface region and cool it, leaving it relatively dark. These magnetic forces can also lift up electrically charged particles from the Sun in huge prominences or flares which can be spectacularly visible, especially at total eclipse. A good-sized prominence weighs as much as a mountain and yet the Sun's magnetic fields are powerful enough to eject this mass upwards at several hundred miles per second. The Sun's magnetic field is being stretched and suddenly released like elastic, whipping a solar flare up into space, escaping the Sun's gravity.

Any flares that shoot towards the Earth make their presence felt. Beautiful auroras are a particularly noticeable sign. The auroras are caused by pieces of atoms – negatively charged electrons and positive protons – which have been ejected by the Sun and then interact with the Earth's magnetic field. This accelerates them; they collide with atoms in our atmosphere and disrupt them with resulting flashes of light.

To see auroras you need clear weather and must live far from the equator, preferably near to the poles. Auroral displays occur most nights in northern Siberia, Lapland, Greenland and Alaska, none of which was overflowing with natural scientists in the seventeenth century. South of this zone are populous regions such as Sweden, Norway and northern Scotland where auroras occur between one per fortnight to as much as three per week on average. Even in London there is one every other month, or 500 in 70 years. During the Maunder Minimum one might have expected

300–1,000 nights of auroral displays in those parts of Europe where astronomers were living. From 1645 to 1715 only 77 were seen. When Edmund Halley saw one on 15 March 1716, it was the first that he had seen – and he was by then 60 years old and had been an avid skywatcher for decades.

If you make a list of the numbers of auroras seen from year to year you will find another interesting message. There is a turn-on in numbers starting in the sixteenth century, with a pronounced pause during the Maunder Minimum before a surge upwards after 1716. Some sensationalists have claimed that this switch-on of the northern lights shows that the Sun has been changing drastically during the last 400 years. However, there is an alternative explanation. The Renaissance in southern Europe during the sixteenth century gave rise to interest in learning and science in particular. But it took time to reach northern Europe, so the growth in auroral observations could mirror the late arrival of the Renaissance in auroral latitudes.

There is no doubt that the Sun does affect us directly, as the auroras show, and this in turn relates to sunspots. Sunspots wax and wane over 11-year cycles. They leave a mark on Earth, the ephemeral auroras, but also make a more permanent record. And by tracing back this record we can tell how the Sun behaved as much as 700 million years ago!

Carbon 14 and the History of the Sun

We are being bombarded continuously by extraterrestrial radiation – cosmic rays. Nuclei of elements produced in distant stars whirl through the magnetic fields of the galaxy, are trapped by the Earth's magnetic arms and hit the upper atmosphere. But first they get blown by the solar wind.

When the Sun is very active, as in sunspot years, it emits flares and the solar wind is strong. This gale shields us from the cosmic rays. Conversely, when the wind is quiet the cosmic rays arrive in hordes. They hit the atmosphere, converting nitrogen into a special form of carbon, called carbon-14. The nuclei of all carbon atoms contain six positively charged protons. Most have six neutrons as well, making a total of 12 constituents in all and called carbon-12. The unstable form, made by cosmic rays in the atmosphere, has six protons as before but *eight* neutrons – a total of fourteen and hence 'carbon-14'. This floats in the atmosphere, mainly in carbon dioxide.

Plant life and trees absorb this carbon dioxide and the carbon decays at a known rate, ending up as the stable form carbon-12. So if you have a piece of wood of known age, you can chemically measure the carbon-14 present now, and work back to infer how much was laid down originally. Tree rings are an ideal source. Each ring represents a year's growth and so the rings in old trees can provide a record of climate, and of the Sun's activity, over several centuries. The proportion of carbon-14 depends on the known age of the ring and the intensity of the cosmic rays that year. There is a pronounced increase in carbon-14 during the reign of the French King Louis XIV, peaking around 1690 – right in the middle of the Maunder Minimum of the sunspot cycle. It is well known in carbon dating circles as the 'De Vries fluctuation'. John Eddy, an astronomer at the Center for Atmospheric Research in Boulder, Colorado, has noticed that the carbon-14 abundance does indeed seem to be related to solar activity. Again this shows up minima and maxima, suggesting that there is a grand cycle of hundreds of years, though it is hard to detect a cycle of a decade because the carbon dioxide settles from the atmosphere over a number of years and its delayed entry into matter smears out the short-term fluctuations.

We can study the carbon-14 content over very long time-scales, back to about 6000 BC. This shows fluctuations superimposed on a grand wave. The carbon-14 was low in the time of the Pharaohs, increasing to a peak in the first 1,000 years AD. It then fell steadily until the early years of the twentieth century, since when there has been a sudden increase. This does not imply that the Sun is to blame; rather that modern society is burning up fossil fuels rapidly and introducing carbon dioxide into the atmosphere, the carbon-12 and carbon-14 content of which contains various mixes over the ages.

The long-term trend – the up and down on a time-scale of

thousands of years – is due to changes in the Earth's magnetic field. The shorter-term fluctuations are most likely due to changes in solar activity which cause changes in the Earth–Sun system and hence changes in the cosmic ray absorption. So there is indeed an 11-year oscillation, which may or may not affect the weather, and longer-term effects, like the Maunder Minimum, which almost certainly do.

An important question is: What is going on inside the Sun that causes these effects? All the solar physicists agree that the solar activity is caused by magnetic forces within it. There are two main theories that build on this.

One theory supposes that the Sun's present magnetic field is what is left over from the time of its formation. In this case the magnetism will be slowly running down and the solar activity with it. If so, the character of its effects on Earth will be changing over long time-scales: even 1,000 years will be too short to reveal this.

The other major contender supposes that there is a dynamo inside the Sun. 'Dynamo' is the general term for a device where mechanical energy is transformed into energy of the magnetic field. The swirling motions of the charged particles sustain the magnetic field over long periods of time. The solar cycle will survive with little change over billions of years. To test time-scales as long as these requires knowledge of geological epochs and some signal of annual activity. Recently George Williams, an Australian geologist, has found out how the Sun was behaving 700 million years ago in Pre-cambrian times.

Near Adelaide in South Australia are the Flinders Mountain Ranges. In a creek bed, lined with eucalyptus trees, are red siltstone and fine sandstone rocks known as the Elatina formation. They are nothing exceptional to look at, but they are important because they carry a coded message about the weather before any life existed. At the end of the ice ages, the amount of the annual floodwater in ancient lakes was greater or lesser depending on the mean temperature. These floods laid down sediments, called varves, which form layers as distinct as tree rings. The Elatina formation is an example.

R. N. Bracewell of Stanford University in California has made a detailed analysis of the varves, which cover a period of 1,337 years. They show clear evidence of 11- and 22-year rhythms modulated by cycles of 314 and 350 years. These tell us about the heat arrival at Earth, and are only indirectly related to the question of the sunspot cycle, but the 11- and 22-year cycles are suggestive.

Now comes the interesting discovery. Sunspots cycle at 11 years on the average but the variation can be as short as 8 or up to 15 years. The internal clock of the Sun seems to put things back on course every 22 years, so that a short span is followed by a long one. Bracewell has found that the cycles in the varve thicknesses also show this sort of behaviour. And it comes about because of the 350-year modulation and its interplay with the 11-year cycle. In the scientific journal *Nature* in 1986 he also showed that the 314-year cycle provides an envelope which determines the strength of the peak activity in the 11-year cycles.

His theory of the four cycles fits the ups and downs of the sunspots remarkably well. He predicts that there will be a steady rise in solar activity until 1991 and that there will be a peak sunspot number of around 100 that year. If he is shown to be right it will prove that the ancient varves are telling us how the Sun is behaving today.

If the Sun's cycle today is much the same as all those years ago then the 'dynamo' theory must be the right one. I await the outcome of Bracewell's prediction for 1991 sunspots with interest. If all goes according to the theory then we can be assured that the Sun is all right; nothing much is changing – at least on the surface. But there *are* hints that something untoward is going on deep in its interior.

7

Is the Sun Still Shining?

Stargazing has always been fascinating. The Sun is the nearest star to us and as such provides a unique opportunity for us to learn in detail how stars shine. This isn't just an academic point, for we rely on the Sun – if it goes out then so do we.

We know how big the Sun is and that it is made of the same sort of stuff as you and me. Pieces ejected during violent flares have blown into our own atmosphere, so although we have yet to visit the Sun, bits of it have visited us. We know its chemical make-up by the dark lines permeating its spectrum; we can see the conditions on its surface, such as its temperature of 6,000 degrees; we know how much heat arrives here and hence how much the Sun puts out. Here on Earth we are desperately trying to satisfy our energy needs while the Sun, each second, is outpouring into space enough energy to keep us going for 1 million years. It has been doing this for 4½ billion years. What is its secret supply?

A burning coal fire releases the energy stored in molecular bonds: chemical energy. The chemical energy emitted by a gram of almost anything is about the same. When you metabolise food your body liberates energy as heat; a chemical explosion gets it over and done with much faster but the net output is about the same.

In the nineteenth century scientists thought that the stars burned much like a conventional fire, different elements combining to form new compounds, giving off heat in the process. This is what happens in a fire: carbon atoms in the coal or wood and oxygen in the air combine and convert into carbon dioxide (smoke) and carbon monoxide (a poisonous gas). This chemical reaction gives off heat.

Gravity is an efficient way of liberating energy. On Earth waterfalls have turned wheels and ground corn for centuries. Today the tumbling cascades of dammed water can generate hydro-electric power enough to light a city. The strong gravitational pull of the Earth's mass is the key and as things weigh 30 times more on the surface of the Sun than they would on the Earth, so is the Sun's gravity more effective as an energy source. It could generate a lot of heat by collapsing under its own weight. A nice analogy that I once heard is that if you dropped a gallon of gasoline into the Sun it would produce 2,000 times as much energy as you would get by burning the fuel.

Stars form when clouds of gas in space fall together under the influence of their mutual gravity. This produces the initial warmth in protostars – fine early on, but stars wouldn't last long if this was the whole story. To generate its heat this way the Sun would have to be using up its fuel at a phenomenal rate to keep us warm at a range of 100 million miles. It would be shrinking by a few tens of metres each year, the distance that a sprinter covers in a few seconds. This is too gradual for us to detect, but go back in time and the Sun would have been much bigger than it is now: 500 million years ago it would have filled the sky. The geological record shows that that cannot be; photosynthesis seems to have been occurring 700 million years ago much as it is now, implying that the sunshine hasn't changed much on that time-scale. Whatever fuels the Sun it is able to do so without significantly altering the Sun's size or its fuel content over hundreds of millions of years.

This puzzle taxed the nineteenth-century scientists. Some anti-evolutionists argued that the solution was obvious: the 'paradox' proved that the Earth was only a few thousand years old, as in Bishop Usher's literal interpretation of the Bible. (In the seventeenth century Bishop Usher had added up the ages of the patriarchs and dated the Earth's creation as precisely 6 p.m. on 22 October 4004 BC.)

Lord Kelvin, one of the leading scientific figures of the time, had addressed the problem too. He also found another conundrum, which concerned the Earth but, as we shall see, bears on the Sun's fuel source as well.

Kelvin had thought about the warm Earth – 300 degrees above absolute zero, isolated in space which is colder than a deep freeze. During the day it absorbs some heat from the Sun but reflects much of the sunlight back into space. By night it is giving up heat rapidly into the coldness of space. Year after year the Earth must have been losing heat: Kelvin calculated that it would have taken no more than 20–40 million years to have cooled from an initial glowing gas ball to its present ambient temperature. Geologists meanwhile were insisting that rocks on Earth were more than 100 million years old. So how come the Earth was still habitable; that it was not already colder than a Siberian winter?

The first clues that there is more going on in the Universe than the Victorians knew came with the rapid series of discoveries on the nature of the atom around the turn of the century. The discovery of X-rays, with their threat to reveal what you had on underneath your clothing, astonished Victorian society. Then in 1896 Henri Becquerel in Paris discovered radioactivity – spontaneous radiation by salts containing the element uranium. Soon Marie Curie and her husband Pierre had isolated an element, radium, whose spontaneous radioactivity is so great that it feels warm and glows in the dark. Here was the evidence for a source of energy that was totally new, something that is beyond mere chemistry. If one could convert the energy to power with 100 per cent efficiency the energy contained within 1 gram of radium would be sufficient to drive a 50 horsepower ship all the way round the world at 30 miles per hour.

Ernest Rutherford and his collaborators in Cambridge and Manchester in England and McGill University in Canada realised that in radioactivity the atoms are exploding; atoms are changing from one variety to another rather than mixing together as in the case of coal or wood burning.

Atoms consist of lightweight electrons surrounding a more massive and compact centre, the atomic nucleus. The atomic nucleus in turn has an inner structure made from bulky particles called protons and neutrons. Chemical reactions involve the peripheral electrons; radioactivity involves the rearrangement of the protons and neutrons in the nucleus.

Here was the promise of unprecedented utopia: atoms of a useless element might be transmuted into those of a more useful one and vast quantities of energy given out in the process. It hasn't quite turned out that way, but there is no doubt that the energy latent in an atomic nucleus is millions of times more than that given up in chemical processes. In a single gram there are so many atoms that they pour out this energy effectively unabated over the ages. Rutherford realised that this was the key to Kelvin's conundrum: radium and other radioactive elements in the Earth's crust provide an extra source of heat that warms up the globe. This extra warmth slows the cooling, with the result that the planet has taken several hundred million years to fall to its present temperature.

This less rapid cooling meant that the planet could live longer into the future as well as being older now. When Rutherford announced this at the Royal Institution in May 1904 the press heralded it as 'Doomsday Postponed'.

Lord Kelvin, proponent of the original idea, was in the audience, which made Rutherford rather uncomfortable. However, soon after Rutherford began speaking, Kelvin fell asleep. Rutherford was worried about the part of the speech where his ideas conflicted with Kelvin's. At last he came to the crucial bit, at which point Kelvin opened one eye and looked at him.

Inspiration arrived on cue, saving Rutherford. He announced that Lord Kelvin had limited the age of the Earth *provided* that no other source of energy was *discovered*. He then attributed to Kelvin that this had been a prophetic remark in that 'tonight we see that radioactivity is the new energy source'. By report Lord Kelvin smiled.

Electrical forces hold the electrons in their remote orbits; much more powerful forces operate within the nucleus and release huge energies when its constituents – the protons and neutrons – are disturbed. We are all too aware of this in the awesome power released in nuclear weapons; it is nuclear reactions that are the cause of the blast.

The atoms of different elements differ in the number of protons in their nucleus. Hydrogen is the simplest and contains but a single proton. Helium, the next simplest, contains two; carbon and oxygen have six and eight respectively. Iron is the most stable configuration of all and contains 28 while the eruptive uranium nucleus contains 92.

Nuclei with more protons than iron's 28 prefer to break up, or 'fission', and release energy. This is the sort of process that occurs in atomic bombs. At the other extreme of light nuclei, such as hydrogen, two or more may join together, building up the nuclei of heavier elements and again releasing energy. This process is called 'fusion'. It is the principle behind the hydrogen bomb and is the source of the Sun's power. So two paradoxes have been solved by the discovery of nuclear energy: the breaking up of heavy elements helps to keep the Earth warm in the refrigerator of space; the fusing together of light elements provides the heat output of the stars.

Atomic Elements

Atoms consist of negatively charged electrons surrounding a central positively charged nucleus; the electrical attractions of opposite charges is what helps to hold an atom together. The simplest atom consists of a single electron and a nucleus with one unit of positive charge. This is an atom of hydrogen. A nucleus with charge +2 attracts two electrons, forming the helium atom. The number of electrons identifies the element.

The number of electrons in the lightest elements and some common heavy ones are listed below:

1 Hydrogen 2 Helium 3 Lithium 4 Beryllium 5 Boron 6 Carbon 7 Nitrogen 8 Oxygen 9 Fluorine 10 Neon 11 Sodium 12 Magnesium 13 Aluminium 14 Silicon 26 Iron 47 Silver 79 Gold 80 Mercury 82 Lead 92 Uranium

The source of power in these nuclear processes is the conversion of matter into energy. Nuclear power stations split the nuclei

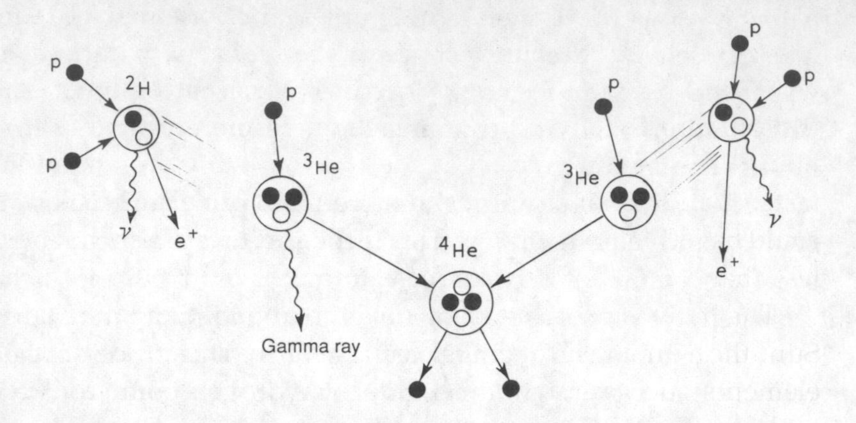

7.1 Hydrogen conversion to helium in the Sun Protons (p, denoted ●) collide and change to neutrons (n, O) by the process $(p) + p \rightarrow (p) + n + e^+ + \nu$, where e^+ is a positron and ν a neutrino which are radiated away. In stage I (top left or top right)

$$p + p \rightarrow p + n + e^+ + \nu$$
$$p + n \rightarrow {}^2\mathrm{H},$$

the proton and neutron having gripped one another tightly to make the nucleus of 'heavy hydrogen' (also called a deuteron), labelled ^{2}H. In stage II (mid-left or mid-right) the deuteron captures a proton and forms a nucleus of helium consisting of two protons and one neutron. Having a total of 3 constituents this is labelled ^{3}He. The process is written

$$p + {}^2\mathrm{H} \rightarrow {}^3\mathrm{He} + \gamma$$

where γ is energy radiated as light.

Stage III (centre). The two ^{3}He nuclei collide, which brings a total of 4 protons and 2 neutrons together. This combination is unstable and fragments instantly into the stable nucleus of helium ^{4}He ($2p$ and $2n$), the remaining two protons being released to initiate the whole cycle again:

$${}^3\mathrm{He} + {}^3\mathrm{He} \rightarrow {}^4\mathrm{He} + p + p.$$

The total chain began with 6 protons and ended up with ^{4}He ($2p$ and $2n$), 2 protons and energy liberated as neutrinos and light. The net change is that 4 hydrogen nuclei (protons) have produced one helium nucleus and radiated energy.

of heavy uranium atoms into lighter nuclei whose combined masses are less than that of the original uranium. The difference in mass appears as energy – the most famous equation in physics,

$E=mc^2$, is at work. If a mass of m grams were totally converted into energy, then an amount E in ergs is produced, where c is the velocity of light (3×10^{10} cm per second). Thus the energy equivalent of 1 gram of anything is 9×10^{20} ergs – some 25 million kilowatt hours. The Sun is producing energy over a million million times faster than this; in each second it loses 5 million tonnes. The Sun could have been radiating energy throughout its life and converted less than 1/5,000 of its total mass into energy.

The process used in atomic power plants cannot operate in the Sun: the Sun consists almost entirely of hydrogen, not of heavy elements such as uranium – 92 per cent of the Sun consists of hydrogen nuclei, 7 per cent is helium and the rest consists of the nuclei of heavier elements, the waste products of the reactor.

The most common occurrence is that after a series of collisions four protons will have fused and produced a single nucleus of the helium atom. This is lighter than the four protons and the spare mass is liberated as energy, ultimately felt as warmth here on Earth.

Protons fuse when they touch, but getting them to do so is like forcing the north poles of two magnets together with a precision of less than a billionth part of a thousandth of a millimetre. To overcome the electrical repulsion the protons must collide at high speed, or equivalently at high temperature. On Earth we can speed protons up in huge accelerators, such as that at the European Centre for Nuclear Research (CERN) in Geneva. The results of such experiments, where protons are fired at one another or at the nuclei of other elements, and other experiments where the energy released in nuclear reactions is measured, are all combined and enable us to deduce the effects of the nuclear reactions that are going on inside stars like the Sun. They imply that, for proton fusion to fuel the Sun, the temperature at its heart must be over 10 million degrees. This thermal agitation throws the protons together; the energy liberated maintains the heat and the conditions for the reactor to continue working.

These are the conclusions from nuclear experiments in laboratories on Earth: 'if the Sun works this way it has to be that hot'. The miracle is that in fact the Sun *is* indeed this hot, and we know this because we know how heavy it is. Its weight causes intense pressure at the centre, some 200 billion times our own atmospheric pressure. The Sun survives collapse because the gas at the centre presses back. Gas under pressure in a fixed volume heats up. From the mass and size of the Sun we can compute the temperature required to prevent collapse. It is 15 million degrees, hot enough to hurl the protons together and fuse them. The various things that we have deduced about the Sun – its mass, the pressure on its inside and the consequent heat, the self-sustaining nuclear fusion under these conditions – all fit neatly together.

So there can be no doubt that this is the source of the Sun's output. Our existence is thus the result of a delicate balance. The fusion happens fast enough that the Sun keeps burning, yet slow enough that it survived long enough for intelligent life to develop on one of its planets.

The conversion of hydrogen into helium is most, but not all, of what cooks inside the Sun. Occasionally three helium nuclei collide and fuse to make a single nucleus of carbon. Now the chance of this would be very small indeed but for the fact that the carbon nucleus happens to vibrate a little with a characteristic or 'resonance' energy. The energy of such a vibrating carbon nucleus is almost identical to the energy that three helium nuclei have and this turns out to make it much easier for the helium nuclei to form the carbon. Were this not so, we would not be here. Indeed, the universe would have very little of anything heavier than helium!

Once formed, the bulky carbon is an easy target for the ubiquitous protons. It absorbs one and turns into nitrogen. Two more hits by protons and a dash of radioactivity produces oxygen, gamma rays and 'hot' ghostly particles called neutrinos which pour out from the Sun into space and can be detected here on Earth. One more hit by a proton and the oxygen splits up into a single nucleus of helium and one of carbon. So we have gone through a cycle: carbon back to carbon with energy radiated in the process. This carbon is like an egg awaiting fertilisation by one of the many protons still in the Sun's centre, ready to go through the cycle again.

The result of all this is that carbon is a catalyst, eating up protons from the Sun's core, changing their spots several times before ending up as carbon and helium and liberating energy. Protons are being used up at the rate of 5 million tonnes every second, each day, year and century through the ages. Slowly the Sun is being changed from a hydrogen bomb into a helium ball. In the meantime, for another 5 billion years or so, there is enough to keep us warm. Unless . . .

Suppose the thermonuclear processes in the heart of the Sun have run down. It could be a long time before news of the changed circumstances in the Sun's core reach Earth's surface. Thus we could be somewhere in this period, waiting for the surface light to be turned off. But the fusions in the Sun's heart have a byproduct – the neutrinos – and their number is like a thermometer: detect them on Earth and you can measure the Sun's internal temperature. A low neutrino flux could be our advance warning of a real energy crisis.

And here we meet the problem: too few neutrinos are arriving compared to the numbers expected if the Sun is behaving as we believe it should. What is wrong? Several people are trying to find out and there is no certain explanation yet but there are several possibilities. Some of these have their own bizarre consequences not just for the Sun but for the future of the entire universe.

First I will describe what these neutrinos are and how we detect their arrival on Earth. Then I will review the ideas that are the current favourites for explaining their shortfall.

The Neutrino Mystery

The neutrino is one of the most pervasive forms of matter in the universe, and also one of the most elusive. Its mass is less than a millionth that of a proton. Indeed, it may weigh nothing at all – no one has yet been able to measure such a trifling amount. It is electrically neutral and can travel through the Earth as easily as a bullet through a bank of fog. As you read this sentence billions of neutrinos are hurtling through your eyeballs as fast as light but unseen. Theorists estimate that there are 100–1,000 neutrinos in every cubic centimetre of space.

An intense but unfelt 'wind' of neutrinos emanating from the nuclear processes in the Sun plays continuously on the Earth. In addition there are lesser breezes of neutrinos from collapsing stars and other catastrophic processes in our galaxy. The Sun is shining in neutrinos with nearly as much power as it shines in visible light and its neutrinos swamp those from other stars as the daytime Sun blinds the other stars.

If our eyes saw neutrinos we would be able to see in the dark. Indeed, there would be perpetual daytime, for the Earth is transparent to the solar neutrinos; there is no neutrino shadow. Neutrinos shine down on our heads by day and up through our beds at night undimmed!

Even the Sun is transparent to these weird particles. Whereas the light produced in the nuclear reactions at the Sun's heart takes 1 million years to fight its way to the surface, the neutrinos stream out in under 2 seconds. When we see the Sun we are seeing the end-product of processes that took place in its centre when *Homo sapiens* first walked the Earth. If we could see neutrinos we could look into the heart of the Sun and see it as it is NOW.

The catch is that you must capture the neutrinos. The difficulty is obvious: if they can escape through the whole maelstrom of the Sun, why should they give themselves up for our convenience? The trick is that, although any one neutrino is extremely unlikely to interact, given a lot of them and enough material waiting for them, occasionally one will be caught. The scale of the problem is illustrated by the fact that a full half-century passed after the Austrian physicist Wolfgang Pauli predicted that neutrinos exist, before anyone captured them and proved the idea to be correct.

Two physicists working at the Los Alamos laboratory in New Mexico in the 1950s were inspired by the idea of 'doing the hardest physics experiment possible'. Clyde Cowan and Fred Reines argued that if neutrinos really existed there ought to be some way of proving it. Neutrinos must do *something*, however rarely, if they have any physical reality. They decided that brute force should do the job: put lots of target in the way of an intense stream of neutrinos.

At first they thought that atomic explosions could provide a suitably intense supply of neutrinos. This is probably true, if rather risky to do as an experiment. Eventually they realised that a nuclear reactor produces neutrinos sufficiently intensely and under controlled conditions. The neutrinos coming from a reactor should occasionally hit atomic nuclei in the target and reveal themselves by changing their spots and changing into more easily detected forms of matter, such as electrons.

Their enormous detection apparatus consisted of 1,000 pounds of water directly in the stream of the neutrinos from the US Atomic Energy Commission's reactor at Savannah River. They named it Project Poltergeist because of their quarry's apparent undetectability. They estimated that the flow of neutrinos from the reactor is about 30 times that expected from the Sun. They succeeded in detecting one or two collisions of neutrinos with the hydrogen nuclei in the water every hour. This was a very feeble signal, but enough to show that the neutrino really exists.

The science of neutrino astronomy arrived in 1964 when Raymond Davis's team from the Brookhaven National Laboratory in New York built a huge detector deep underground. They had to do this in order to hide from all the other debris that continuously hits the upper atmosphere – the ubiquitous cosmic rays. These are mostly atomic nuclei, produced in violent processes deep in space and captured by the Earth's magnetic field. The atmosphere shields us from these as effectively as from other larger invaders – no more than a gentle drizzle of atomic particles reaches ground level, and very few get through a few *metres* of earth. Go down a deep mine and neutrinos are the only survivors.

So not only are neutrinos hard to capture but you have to go to extreme lengths – or rather depths – to eliminate the competition.

Davis's team went down a full mile in the Homestake Mine at romantically named Deadwood Gulch, South Dakota. The mine was too small for the gargantuan device they had in mind. They first had to remove 7,000 tonnes of rock, then the tank had to be built on the surface and lowered bit by bit to the mine. Once they had it all assembled deep underground they filled it with over 100 tonnes of perchlorethylene – dry cleaning fluid!

They chose this common liquid as it contains a lot of chlorine atoms. When a neutrino from the Sun penetrates into the chamber and hits a chlorine atom it has a good chance of changing the chlorine into argon. Periodically Davis flushed the argon from the tank and measured how much there was. Knowing how much argon is being formed tells you how many neutrinos have hit.

They started the experiment in 1970 and have repeated it some 60 times. They are seeing the Sun shining in neutrinos, but it is dimmer than expected. On the average only one-third as many neutrinos arrive in their detector as ought to if our understanding of the Sun is correct. In addition there are small fluctuations in the neutrino flux, which some people claim follow the pattern of the sunspot cycle.

In a talk about the neutrino experiment at an American Institute of Physics conference in 1984, Ray Davis's group described how the neutrino flux varied year by year from 1970 onwards. Each year they have a datum that is uncertain by some amount owing to the many complexities of the experiment and trying to draw conclusions from a small sample of neutrino hits. If you ignore these uncertainties, there appears to be an up and down that correlates with the sunspot cycle; if you take the uncertainties into account, however, the correlation is much less impressive.

J. Bahcall of Princeton, one of the leading theoreticians, has made a detailed study with G. Field and W. Press of Harvard to see if there is a real correlation or not. He finds that the data are 'suggestive but not statistically significant'. The suggestive correlation stems almost entirely from a low neutrino flux near the start of 1980 during a sunspot maximum. The correlation may be insignificant but interest has been aroused and so we await the next sunspot maximum to see if the neutrino flux drops again.

Something untoward is happening and we don't yet know what it is. New experiments are under way trying to solve the puzzle. To appreciate what new insights they will offer we need to review what burns inside the Sun according to present wisdom.

Arizona meteor crater with San Francisco Mountains in the distance.

Photos © Meteor Crater Enterprises

Close up of the crater with tourists setting the scale.

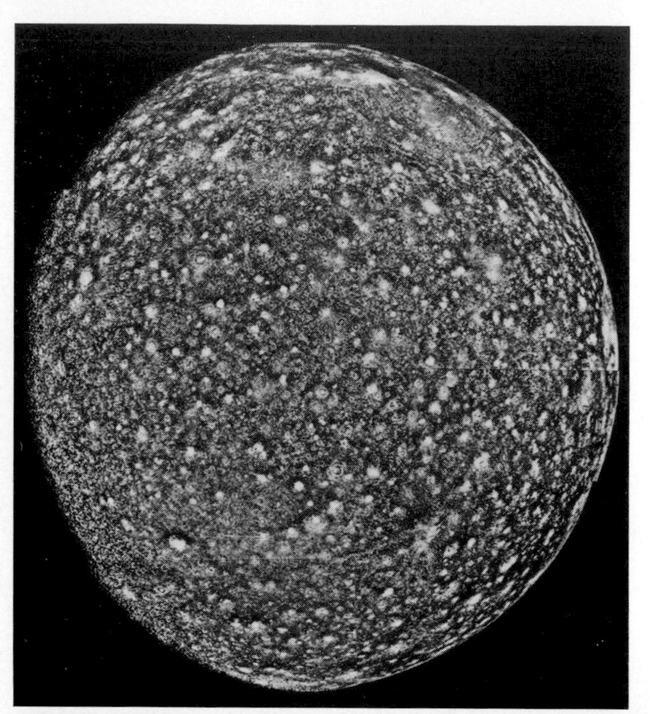

Craters on Callisto
Meteorite impacts have cratered the entire surface of Callisto, one of Jupiter's satellites. A close up view of the region at top right (*shown below*) reveals a huge impact basin over 350 miles diameter. The concentric rings are the result of the icy crest responding to shock waves produced by the large impact. They extend over 600 miles.
Photos © NASA/Science Photo Library

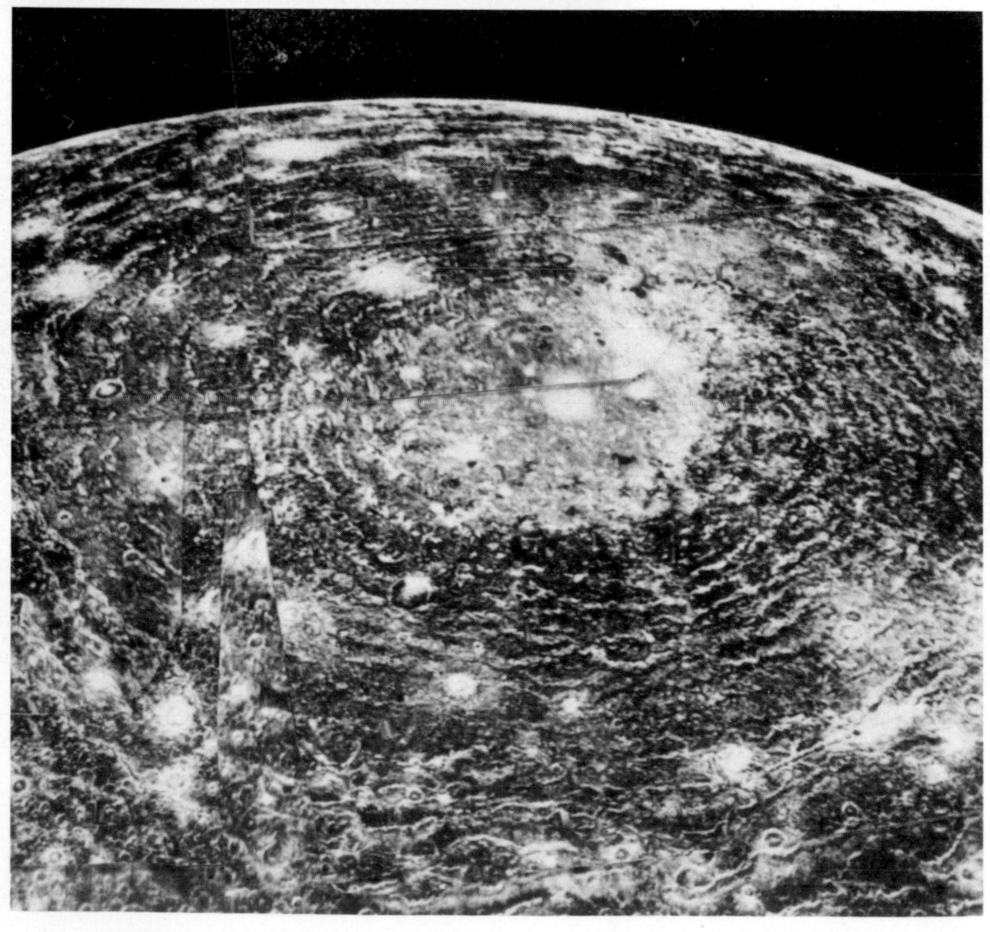

Epimetheus (*top*). Saturn's eleventh satellite is shaped like a tooth 80 miles by 40 miles, and is probably the fragment of a violent collision. Similarly Saturn's satellite **Hyperion** (*bottom*) whose shape has been compared to a peanut 200 × 100 miles across.

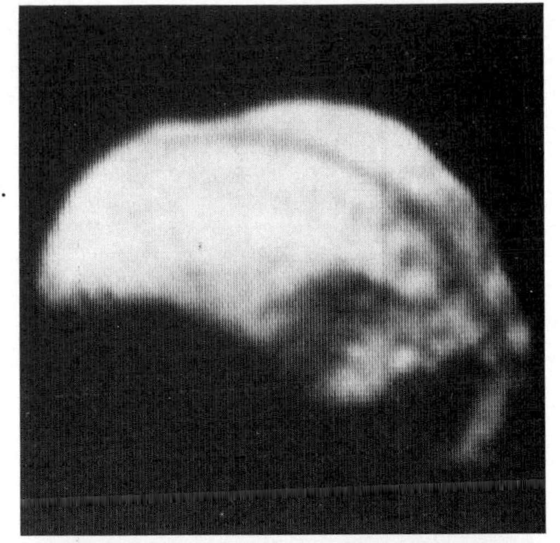

Mimas (*centre*). The Herschel crater on Mimas, one of Saturn's satellites, is 80 miles diameter, 7 miles deep with a central peak that is 4 miles high; it was probably caused by impact with a body about 8 miles wide. It fills about 1/3 of Mimas' size. An impact with a body only slightly larger would have smashed Mimas into several pieces.

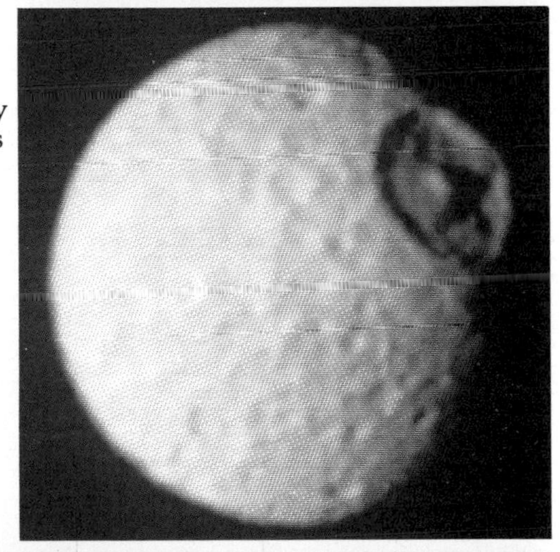

(*Above*) **A southern hemisphere view into the heart of the galaxy.** The trail is due to a meteorite which may have landed in the Australian bush.
© Fitzharries Astronomical Society

(*Left*) **Comet West**, photographed in 1976, the year of its discovery. This bright comet will not return for thousands of years.
© Betty Milor/Science Photo Library

(*Top of opposite page*). **Our Galaxy In Space.** This shows how we would appear to observers in remote galaxies. The Milky Way is half way up on the extreme right. The faint galaxies near to us are our satellites, The Large and Small Magellanic Clouds. Below us are two major galaxies, Andromeda and M33. This cluster of galaxies forms the "Local Group" which in turn form part of the Local Supercluster of galaxies centred on the Virgo Cluster (extreme left).

Milky Way in Space (*see caption opposite*) © *David Parker/Science Photo Library*

Cannibal galaxies; NGC 7752/3. Gravity is pulling stars from the small galaxy into its large neighbour. © *Royal Greenwich Observatory/Science Photo Library*

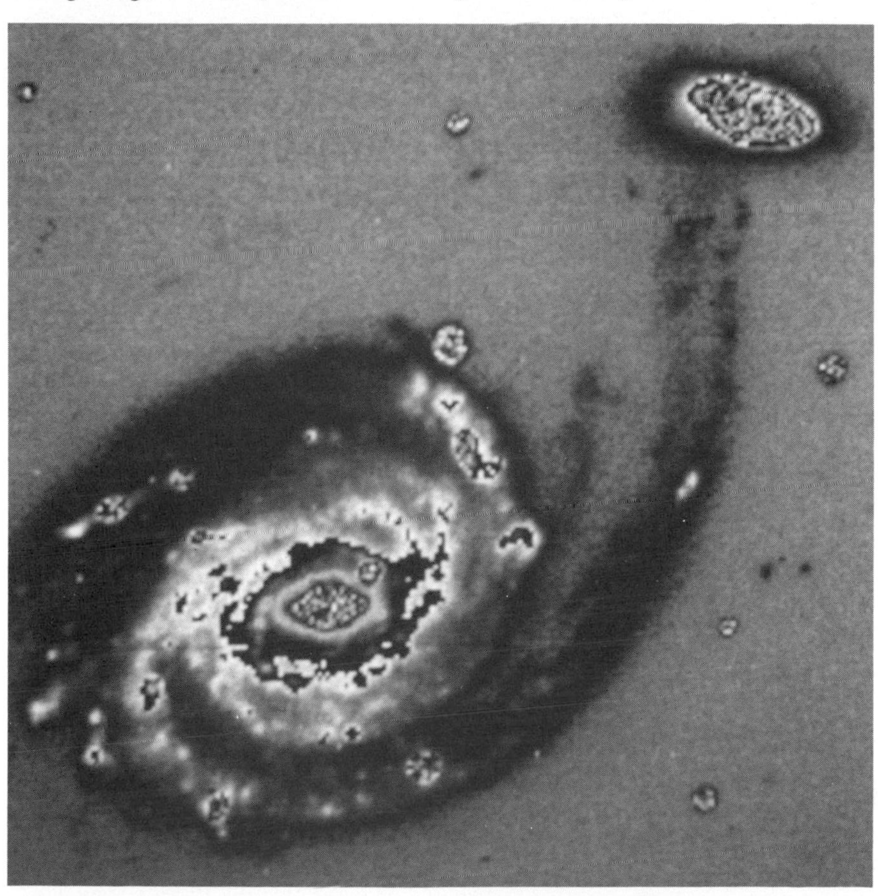

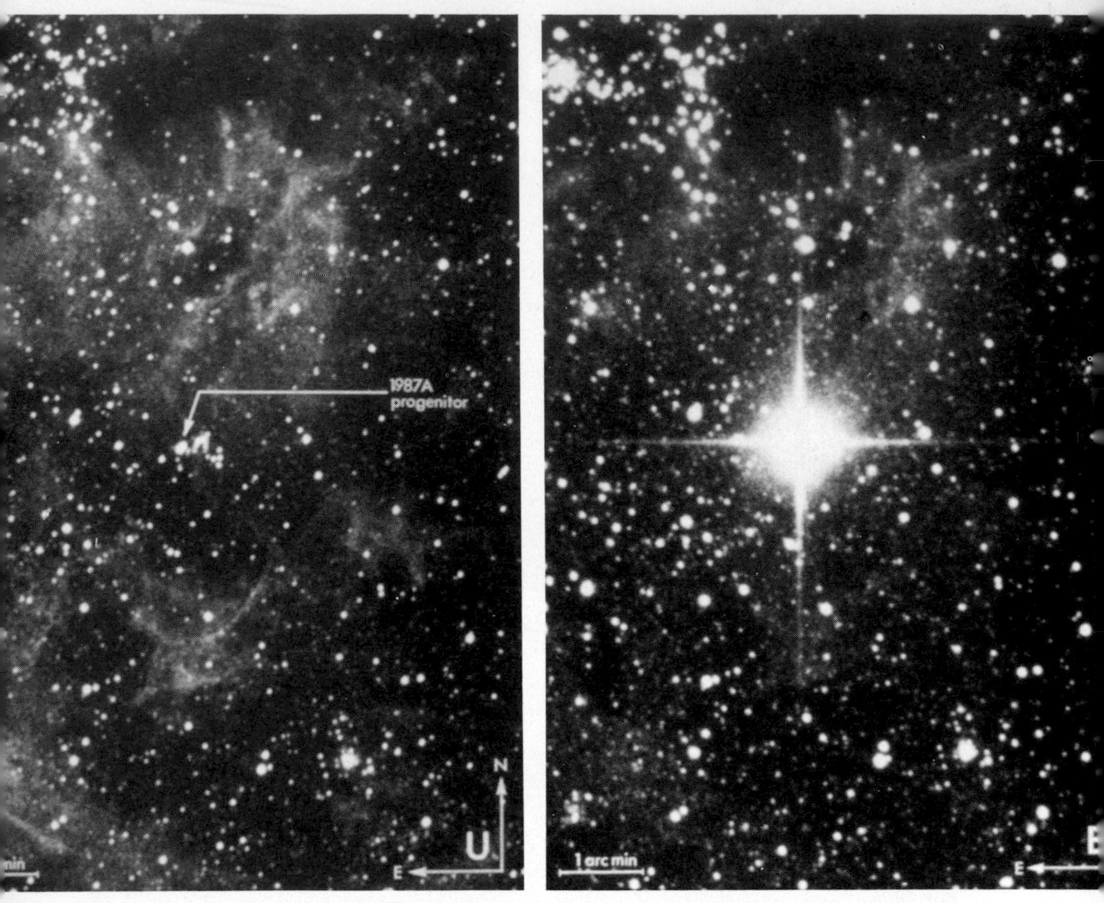

Supernova in the Large Magellanic Cloud The Large Magellanic Cloud became famous in 1987 as the home of the first naked eye supernova for four centuries. These two images show the supernova and the surrounding stars photographed with the ESO Schmidt telescope on Feb 26, 1987 (The cross is an optical effect in the telescope). The surrounding stars can be identified in the left hand picture which shows the scene before the central star exploded; the picture on the right shows the same star field after the supernova explosion.
Photos © European Southern Observatory

(*Opposite page; top*) **The Crab Nebula** is the ejected remains of the supernova of AD 1054. They are moving away from the centre at 1000 miles per second, so fast that the shape changes in a human lifetime. A pulsar, neutron star, has been detected at the centre of the Crab Nebula.

(*Opposite page, bottom*) **The Ring Nebula in Lyra**, M57, is a shell of gas thrown off by the explosion of a dying star 5500 years ago. The ring is expanding at 12 miles per second and comes from the star at its centre which is probably a white dwarf.

Crab Nebula (*see caption opposite*)
© Hale Observatories/Science Photo Library

Ring Nebula in Lyra (*see caption opposite*)
© Lick Observatory/Science Photo Library

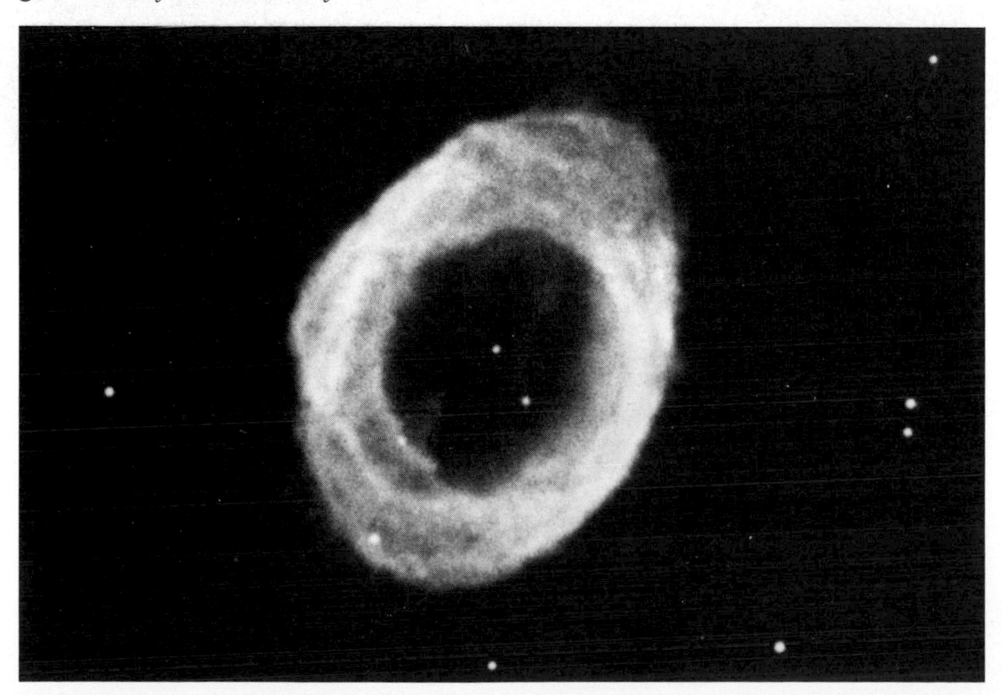

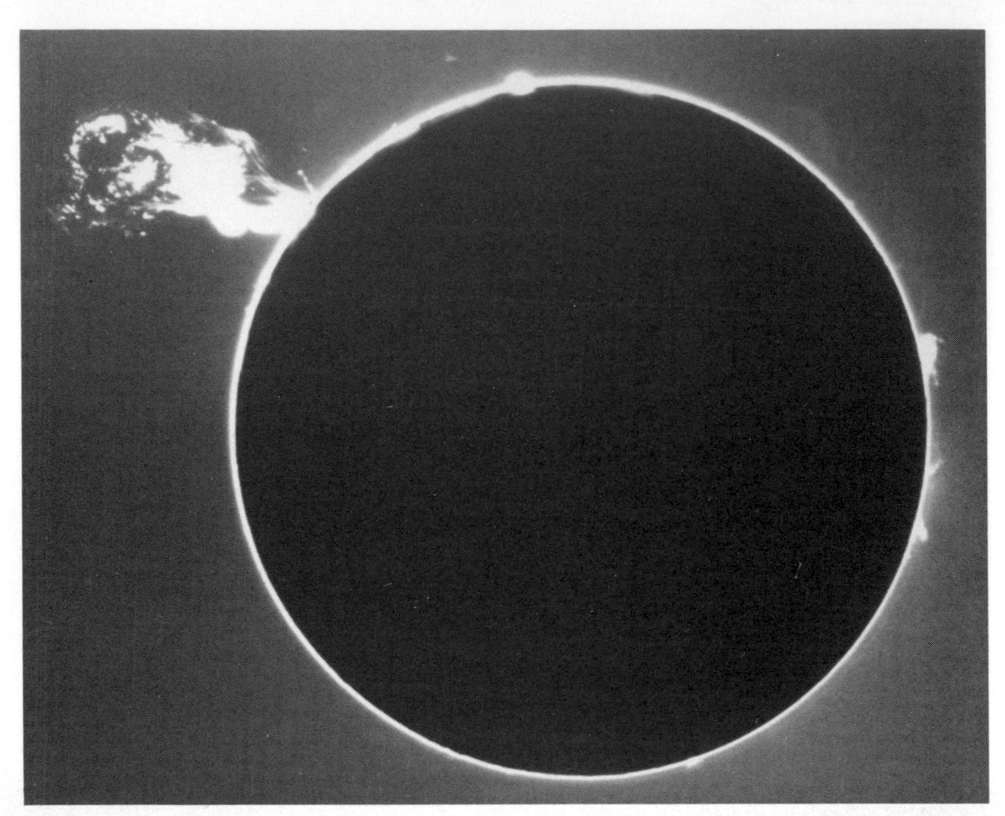

A massive solar flare photographed in 1974 by Robert Fisher of the Halekala Observatory on Maui, Hawaii

Giant solar prominence
Photos © High Altitude Observatory of National Center for Atmospheric Research, Colorado/SPL

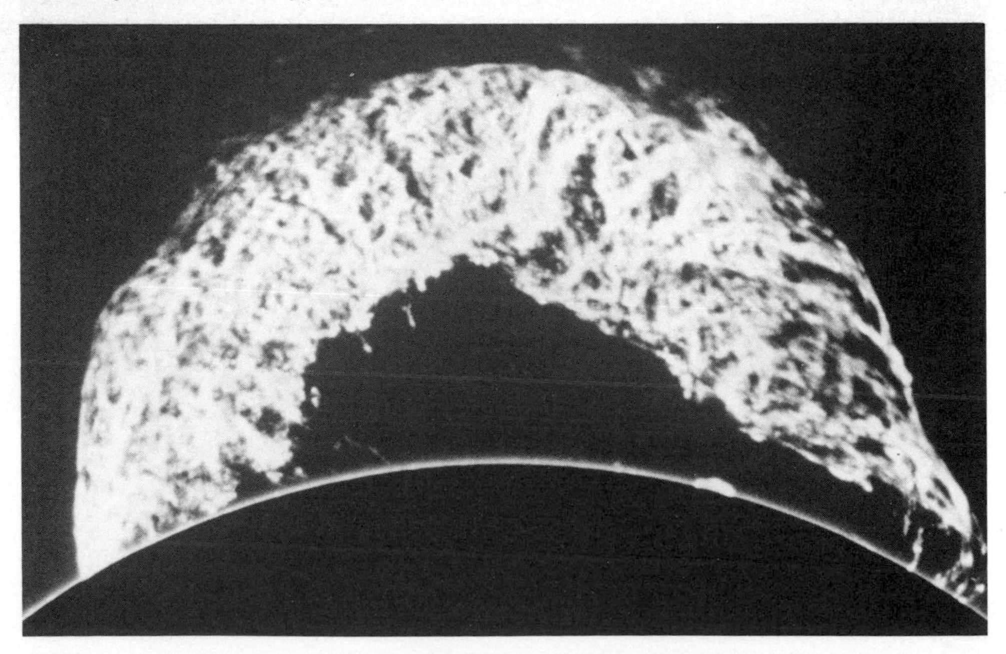

Inside the Sun

The simplest stable particles, the electrons and protons, were formed in the heat of the Big Bang. Radioactivity can transmute protons and their electrically uncharged counterparts, neutrons, back and forth one into the other and help build up the nuclei of heavy elements. Great heat is required to bring the various ingredients together and cook them like this. Gravity provides the fuel in that it tugs the ubiquitous protons (nuclei of hydrogen) together, heating them up into conglomerates that we call stars. It is by means of the stars that Nature can cook the elements that are necessary for the complexities of life.

I described on p. 92 how the Sun does this. Let's look at this again but concentrate this time on the 'waste' products – the radiant energy.

The path to you and me began with the meeting of two protons in the centre of the Sun billions of years ago. They fused, one of them turning into a neutron, and in combination they formed a stable system, a nucleus of 'heavy hydrogen'. A single proton and a single neutron can bind tightly and survive; their combined masses are less than those of the two protons and the spare energy is radiated, in part by neutrinos. These neutrinos can carry off anything up to 420,000 electronvolts of energy (a 1 volt battery could give a single electron an energy of 1 electron-volt, or ev for short). This is the most important neutrino producer in the Sun – over 99 per cent of solar neutrinos are produced this way.

Another possibility in the hot dense conditions of the Sun's centre is that three particles meet. Two protons and an electron can fuse to make the same nucleus of heavy hydrogen as before but this time releasing more energy – 1.4 MILLION eV – again carried away by a neutrino. These nuclei of heavy hydrogen soon are hit by further protons, building up nuclei of the next simplest element, helium. Again, energy is liberated as the protons cluster into ever more stable configurations, but in the building of helium the energy comes out as gamma rays, a high-energy form of light. Collisions among the helium nuclei produce small quantities of the next lightest elements: lithium, beryllium and boron. In this latter stage a few neutrinos are produced with energies as high as 14 million eV (14 MeV for short), but this is 10,000 times as rare as the lower-energy neutrinos produced in the initiating reactions above.

The basic concentrate of reactions in the Sun is producing neutrinos whose individual energies are low but in such vast quantities that they are the main energy carriers in total. The experiment on Earth which has a detector of cleaning fluid sees *none* of these; it is able to capture only neutrinos whose energies are *more* than 1 million eV, and the vast majority of the solar neutrinos have energies that are far below this threshold – recall that over 99 per cent of the neutrinos have energies less than 420,000 eV.

The neutrinos that we are detecting at present are produced by a relatively insignificant process where the elements boron and beryllium are being transmuted in the Sun. The important neutrinos from the proton thermonuclear fusion are out of the reach of present detectors. It is as if the detector's ear is sensitive to notes in the treble clef whereas the Sun is sounding primarily in the double bass. To 'hear' these, the largest part of the orchestra, new types of detector must be built.

You can detect these low-energy neutrinos if you use a target consisting of the element gallium instead of chlorine (cleaning fluid). Gallium responds to impacts from neutrinos having as little as a quarter of the minimum energy needed to affect the cleaning fluid. This will enable us to 'listen' to the Sun's main theme tune. The problem is that to do so we need many tonnes of gallium, which is larger than the annual world production. However, even a modest effort here would be better than nothing and two groups are developing such an experiment. One is in the USSR and the other is an international collaboration that will hide underground in the Gran Sasso tunnel beneath the Alps.

A group from Oxford and a European collaboration are each attempting to develop detectors comprising the rare element indium, which will be sensitive to neutrinos as low as 150,000 eV – a 50 per cent improvement even on the gallium detector – and which, if it works, will pick up the neutrinos from the Sun's most important thermonuclear reaction.

So it is possible that experiments so far, which captured fewer neutrinos than had been expected, are merely picking up the whisper of the Sun's higher register, which is a bit quieter than we expected, while the deep bass is playing fortissimo but unheard as yet.

In a few years when the new experiments begin we may know if this is the answer to the puzzle. Many astrophysicists doubt that the solution is that easy; they believe that we will find a shortfall in those neutrinos too. If they are right, then there is a genuine problem. So we must look at two suspects. Is it the Sun's centre that is cool, or does something happen to the neutrinos en route?

Suspect Number 1: A Cooler Sun

The reaction of some theorists who believe strongly that they understand the workings of the Sun is to declare that the experiment must be wrong. Experiments like these are complicated and sophisticated. Dry cleaning fluid sounds like a kitchen sink enterprise, but don't be fooled, this readily available commodity is surrounded by highly detailed technology. Somewhere in the chain of electronics and computer logistics something might be going wrong, misleading us into thinking that fewer neutrinos arrive than in fact do.

The experimentalists respond that they have checked their apparatus in detail. They have performed a series of tests of the large-scale chemistry, verifying their efficiency at extracting the crucial argon from the chlorine target. Anomalous results when the neutrinos come from the Sun do seem to imply that something is happening in the Sun, or to the neutrinos en route, and not that there is a fault in the apparatus.

So perhaps the standard model of the Sun is wrong.

The neutrinos picked up in the Dakota mine are the end-product of a series of fusion reactions that are driven by heat. It is this heat that throws the protons together; the hotter or colder the Sun is, so the faster or slower the fusion occurs. The rate of neutrino production depends very sensitively on the temperature, varying in proportion to the temperature multiplied by itself 10 times. This makes me nervous: if the temperature is only 10 per cent less than your theory expects (and that doesn't seem unreasonable) the neutrino flux will be cut back by more than a factor of two. Even so, some theorists say that they are confident in their temperature predictions to this level of precision and that this 'easy' way out is not the right one.

However, these calculations implicitly assume that the heart of the Sun consists of protons and neutrons – constituents of the forms of matter familiar to us – and that there are no hitherto unknown particles in there. Which brings us to the modern theories of matter, which some theorists believe could radically alter our picture of the solar furnace.

These new theories suggest that there may be very massive stable particles around, produced at the Big Bang and very rare today. They interact with matter very weakly and are known as WIMPS – Weakly Interacting Massive Particles. Experiments are in progress around the world looking for such things, but there is no direct evidence for them yet. One idea is that, being massive, any on the Earth will sink into the centre under the influence of gravity. In turn, any in the solar system will tend to gravitate into the Sun, and accumulate in its heart.

The standard theory of the Sun implies that its temperature rises rapidly as you near the centre. WIMPS can smooth out this temperature rise, effectively cooling the Sun's heat. To do this the WIMPS have to be between 5 and 20 times as massive as the ubiquitous protons that are the fuel. Lighter than this and they evaporate out from the centre; heavier and they sink there for good. But in the middle mass range they sink in, circulate and carry heat away from the centre as they depart, returning inwards again in perpetual orbit. With this lower central temperature, the fuel cycle is slowed down and the flux of neutrinos reduces in line with observation.

Or at least the neutrinos that the cleaning fluid detects will be reduced. Although the WIMPS are sluggish they could bump into one another occasionally and mutually destroy themselves. A byproduct of this cataclysm will be neutrinos whose individual energies are thousands of times greater than those from proton fusion – the main fuel supply. If the energies of the neutrinos in solar fusion are compared to the sounding of the double bass, then WIMPS give neutrinos a high pitch like a dog whistle. Taking this analogy further: present solar neutrino experiments are sensitive only to a limited range in the treble clef. A search for such a high-pitched (or high-energy) signal is now underway but it will be some time before the results are known.

Suspect Number 2: Neutrinos Are Fooling Us

The smart money lays the blame for the solar neutrino problem on the neutrinos themselves. If neutrinos are massive they can change their character en route from the Sun in such a way that one only detects about one-third of those that set out. Then everyone could be happy: the flux of neutrinos setting out from the Sun would be exactly as theorists predict; the Sun would be shining to order – and the universe will eventually collapse under the weight of these neutrinos!

First things first!

Three varieties of neutrino are known. The processes in the Sun produce one of these (known as the 'electron-neutrino'). If this variety of neutrino is massless it will stream across space to the waiting detector in pristine condition. The flux recorded in Dakota will be a direct measure of that put out by the Sun. However, if the neutrino has a mass, it can change into one of the other two varieties en route. As the detector only records the arrival of the electron-neutrino variety, it will record fewer than set out.

But do neutrinos have masses or not? There is no known principle that says that they must be massless. There is even an experiment performed in Moscow that claims to have measured a small mass, but as this has not yet been independently reproduced current scientific opinion is that the question is still open.

Everyone agrees that if neutrinos have masses, they are very small. The best experiments in laboratories on Earth imply that the electron-neutrino can weigh in at no more than one-thirtieth of a millionth the mass of a proton. People are looking at the flux of neutrinos coming from reactors and accelerators on Earth to see if the flux dies off with distance as the electron-neutrinos change their form. Some groups claim an effect; others dispute this. This is an active area of research at present. The whole nature of neutrinos is one of the great mysteries of matter that the particle physicists are trying to unravel. The problem of the Sun highlighted the conundrum.

It is possible that the neutrinos are changing their identities in flight across space, but recently several theorists have looked back nearer to the start of the neutrinos' journey. Produced in the solar core, the neutrinos have first to shoot through half a million miles of solar material. Could it be that the solar environment is affecting them before they emerge into the light of day?

In 1986 Hans Bethe, famed for unravelling the cycle that produces heavy elements in the stars, extended and popularised an idea of the Soviet physicists Mikhaev and Smirnov. They had pointed out that the propagation of neutrinos through matter, as in the Sun, could have remarkable effects and explain the solar neutrino problem.

As light can become distorted as it propagates through materials with differing refractive indices, so can a beam of massive neutrinos perform funny tricks when it passes through matter of varying density. The Sun is a perfect example of such an environment. The Sun puts all its energy into one type of neutrino (the electron-neutrino) but a resonance phenomenon as this variety of neutrino passes through the Sun's gases can cause the energy suddenly to drain into another variety. To describe how this works would be of little help, but there is a simple analogue that you can build and is a fun party trick.

With a long piece of string and two weights make two swings suspended from a bar. Tether one end of the string and attach the other end to a handle so that you can vary the length of one swing. Start off with the swings having different length and set ONE swinging.

All the energy is in one swing; this is like the neutrino put out by the Sun – all the energy is in this one.

Now change the length of the swing *very gradually* so that the lengths of the two swings get more and more similar. You will find that as the lengths coincide, suddenly one swing will cease and the other will take up the motion. Without touching them, all the energy will have gone across from the initial swing into the other one. The changing length of the swings is the analogue of changing density in the Sun. At a critical condition (in the case of swings it is the equal lengths) one neutrino gives up the ghost and the other takes over.

If this is the explanation of the solar neutrino problem it works because the Sun has a certain density profile AND because the neutrinos are emitted in a certain critical energy range. Consequently the experiments now beginning, which will be sensitive to the lower-energy neutrinos, will see very different effects from those done so far. The science of neutrino astronomy is now beginning in earnest and promises to be fascinating. I suspect that our understanding of stars will prove to be still primitive.

A Link with Sunspots?

Finally, there is the tantalising suggestion that the rate of neutrino detection by Ray Davis's experiment varies with the sunspot cycle.

The Sun responds only slowly to changes, so the neutrino production at the core should be constant for periods in the range 10,000 to 1 million years. It would be a surprise, then, if variations in neutrino flux on a period of the sunspot cycle – a mere handful of years – were due to changes in the core. However, as Lev Okun and a group of Soviet theoreticians have pointed out, if neutrinos interact with magnetic fields then a correlation with the solar cycle should happen.

Neutrinos spin as they fly. The earthbound detector can only detect neutrinos spinning in one particular direction – which is fine if they're not magnetic but disastrous if they are. If they are magnetic, the magnetic field in space will perturb them in flight and change their orientation, with the result that not all will be detected on Earth. The flux will appear to have dimmed. Moreover, the neutrino flux would also be affected by the Sun's magnetic cycle. The sunspots are the outward sign of magnetic activity in the Sun, which varies over an 11-year cycle. If the neutrino is magnetic, then the flux of neutrinos arriving at Earth should vary in phase with the sunspot cycle.

The Sun is spinning, so it has a north and south hemisphere as we do on Earth, and the direction of its magnetic field is opposite in the two – like the north and south poles of a magnet. The Sun's equator is tilted relative to our orbit, so in one half of the year we look more at the Sun's south pole and six months later we're peering in from the north.

If neutrinos are affected by magnetic fields, those streaming out of the northern hemisphere will have been affected differently from those emerging from the southern hemisphere. There is a slit in the Sun's magnetic field along its equator (which is why no spots are seen in this region) and so neutrinos coming out in equatorial regions won't be affected at all by magnetic forces.

The Earth's orbit crosses the solar equator in June and December, so neutrinos that hit us in mid-summer and winter aren't affected by magnetism. However, in spring and autumn of years when the Sun is active the neutrino flux should be affected if these ideas are right. So there may be a six-month variation in the neutrino flux as well as a larger-scale 11–22-year cycle.

There may be signs of this in the data but it is not significant statistically. (If a coin lands heads four times in a row that is chance; 40 times and someone must have loaded the coin – the solar data so far are like the former.) Whether the effect is there or not all hangs on 1980 when solar activity peaked and the solar flux dipped. Ignore that one year and there is no effect visible over several years. But the question has been raised: will it happen again when solar activity peaks (as we expect) in the early 1990s? If it does it will teach us about the neutrinos – they feel magnetism – and the variation will be real. However, this is up and down on top of an overall depletion. The cause of the average depletion may still remain an open question.

In the next few years studies of solar neutrinos promise to teach us a lot about neutrinos, about the Sun and maybe even about the fate of the universe. If any of these explanations is the reason for the anomalies then it is likely that neutrinos have a mass. If Bethe,

Mikhaev and Smirnov are right, then neutrinos are affected inside the Sun not outside, and it is unlikely that we will be able to measure their mass on Earth. But we will be able to test the idea when the new detectors of the low-energy neutrinos, those produced in the core, come into operation. According to Bethe these neutrinos will be affected very differently from the ones we've seen so far. In a few years we may know if this is the answer. If so, the neutrino mass is very small and will have only a marginal effect on the gravity of the Universe.

However, if the neutrino's behaviour in flight is the reason, then neutrino masses could be large enough that they will significantly control the future evolution of the universe. There are so many of them around that they could provide more mass than all of the stars that we have ever seen through the most powerful telescopes. The gravity of these neutrinos could make the universe collapse in a big crunch.

If this is the future, the collapse is still many billions of years away but it would be an incredible achievement of the human intellect if a careful study of our Sun and the solar neutrino problem led us to foretell the eventual fate of the entire Universe.

Part III

A Galaxy of Stars

8

A Journey around the Milky Way

The Scale of Things

Two policemen were driving across the desert in California one night 30 years ago when they saw a flying saucer. Or at least, that is what they reported.

Hovering over the road, directly in front of them, was a bright light. In their report of the event they described it as 'blinding'. At first they thought that it was the landing light of a plane from a nearby airforce base, but it hovered, absolutely still, and then suddenly shot away into the distance at 'incredible speed'. As UFOs were then the rage, they decided that the light must be from another world.

They were right! The light indeed came from another world – the planet Venus. On some nights if the conditions are clear and Venus is low in the sky it can appear as bright as a car headlight. Once you mistakenly interpret the light as being nearby, the dimming by a passing wisp of cloud gives the impression of sudden departure. Many reports of UFOs turn out to be sightings of Venus.

Not only Venus but Mars and Jupiter also can shine brighter than any star. Yet none of the three, nor indeed any of the planets, shines of their own accord. Unlike the stars which emit their own light, much like the Sun, the planets are mirrors. We see sunlight reflecting from their atmospheres or surfaces.

These planets reflect only a minute portion of our Sun's light. Yet they outshine other stars completely. Ponder this fact for a few moments and you will begin to get a feeling for how remote those twinkling stars really must be. The distances involved are so vast that it isn't helpful to express them in miles. Do you have a feeling for the difference between a billion billion billion billion and a million billion billion billion billion? They are both huge, beyond comprehension. So astronomers use another measure – the time that light takes to get from 'there' to 'here'. Even these can be confusing unless you relate them to more familiar time-spans.

Light is the fastest mover in the universe. It travels 300 metres in only one-millionth of a second. Human reaction times are a few thousandths of a second. On the freeway your car will have travelled a few metres before you react to an emergency. Light will travel 1,000 miles in this time. People in the north of England receive their radio time-signal from London delayed by one-thousandth of a second. The message in a long-distance telephone call between Glasgow and London takes a similar time. We don't notice such brief delays; they are 'instantaneous' for us. When you make a transatlantic call the signal may be routed up to a satellite far above the Earth, and back down again to the other party. The delay of up to ½ second can be quite disconcerting. You are beginning to be affected by the finite speed of light and radio signals.

We see the Moon as it was 1.5 seconds ago – the time of two heartbeats. The Sun is more remote. You can cycle 1 mile in less time than it takes sunlight to reach us. The Sun is '8 light minutes' distant.

Now that we are sending space probes to the outer reaches of the solar system, we are beginning to feel the 'light lag' directly. Good athletes complete a marathon in the time that light travels out to Uranus. This created real problems for scientists when the Voyager probe arrived there in 1986. Voyager passed through the rings and moons of Uranus, preprogrammed to take pictures and transmit them back to Earth. For 3 hours the signal rushed across space towards the waiting scientists.

At last the first exciting pictures emerged. It was no use anyone asking 'Can you tell the probe to point its camera back so that we can have a closer look at that new moon?' The moment anything new turned up, it was already too late. It would take 3 hours for a message to reach Voyager, by which time it would have left Uranus far behind.

It takes half a day for light to depart the solar system. The nearest stars, Proxima Centauri and Alpha Centauri, are nearly 4 years distant. These are visible in southern skies. Northern viewers can see bright Sirius shining as it was 8 years ago.

When we look at the stars we are looking back in time. We see Betelgeuse, the red star in Orion, as it was at the Norman Conquest, the bright nebula of new stars in Orion's sword in the early Christian era, and the Crab Nebula in the time of the Egyptian pharaohs.

Our galaxy consists of 1 billion suns, most of which are in a thin disc with a bulge at the centre. When you see the Milky Way arching overhead you are looking through the plane of the disc. If you live in the southern hemisphere you can follow the Milky Way to Sagittarius where you will be looking right into the central bulge the heart of the galaxy. There are so many stars there, so far off, that individual stars cannot be distinguished; the galaxy's centre appears as a pool of light. That light has been travelling 33,000 years to reach us.

All the stars in the constellations are in our galaxy but appear thinly spaced because we are looking out of the plane where most of the stars lie. On a clear night you will also be able to make out some nebulous objects that look like thin misty clouds reflecting moonlight. Some of these are not in our galaxy. In southern skies there are the Magellanic Clouds, visible to the naked eye and named after the exploring seafarer Magellan. These are two galaxies, smaller than ours, satellites of the Milky Way. Trapped by our galaxy's huge gravity, they orbit around us continuously. They are nearly 200,000 light years distant; we see them as they were when Neolithic man occupied the Earth.

In northern skies you can look at the constellation Andromeda and see a faint wisp near its second brightest members. Through a telescope this is a marvellous sight. It is a huge galaxy much like our own, where the stars appear in spirals (Figure 4.2). It is 2 million light years away; its light has been travelling as long as mankind has existed. If we could see the Milky Way from outside it would look much like the Andromeda Nebula does from here. Viewed from a planet orbiting a star in the Magellanic Clouds, our entire Milky Way would be a brilliant vision in the heavens.

Here we begin to be aware of the clumpy nature of the Universe. The Sun is our nearest star and the entire solar system surrounding it occupies distances that light can travel in a single morning. The next star is 3,000 times more remote than this, a light voyage of 4 years. Stars occur on average some 4 or 5 light years apart throughout the galaxy, extending 100,000 light years from edge to edge. Then vast darkness again until we find our nearest galactic neighbours, the Magellanic Clouds 200,000 light years away. Then there is nothing but the vastness of space, dark gases until the next major galaxy of stars, Andromeda, 2 million light years away, ten times more remote than our satellite galaxies. The nearest quasars (quasi stellar objects) are 10,000 million years distant. This is over half way to the start of time, for the Universe was born at the moment of the Big Bang, some 20,000 million years ago.

Viewed from the Earth our Sun is very bright because it is so near to us, but place it in a constellation and it wouldn't be visible; it is an unremarkable boring star. Viewed from the Magellanic Cloud, looking into our galaxy from outside, a creature wouldn't notice the Sun either. Suppose that we were looking at the Milky Way galaxy from outside; whereabouts would the Sun be?

The most noticeable feature of the galaxy is the bright central bulge. But we don't live there. Radiating out from the centre are dense spiral arms (see Figure 4.2, page 48). We don't live there either but you're getting warm. Once upon a time it was thought that the Earth was at the centre of the solar system. Now we know that is not so and that the Sun isn't even at the centre of the galaxy. Our galaxy isn't anything very special either: encircled by two minor galaxies – the Magellanic Clouds – and replicated throughout the cosmos, Andromeda being distinguished merely by its proximity.

There are two ways to hide and the most effective is anonymity – lost in the crowd. The stars in the galaxy are as numerous as grains of sand on a beach. We are one of millions of unremarkable points out in the suburbs. That's us – the X in Figure 4.2. No one living in the Andromeda galaxy would give our sun a second glance, even supposing that they had seen it! How many existences are out there orbiting a star that we haven't even noticed?

It is impossible to stay at rest when you are being tugged by the gravity of all these stars and galaxies. All stars in one galaxy collectively tug on those of a neighbouring galaxy. Whole galaxies are on the move. Over the aeons galaxies will disrupt one another, which will affect the long-term future of the Universe. Of more immediate concern is the behaviour of the stars within our own galaxy, and the ones near us right now.

Just as the galaxies are in motion so are the stars within. As the Sun and Moon make tides within the Earth, so do nearby galaxies make tides within our own galaxy. The stars are tugged this way and that over the aeons. Our galaxy is rotating like a huge Catherine wheel. This fluid of stars is turbulent; within the general flow there is chaotic motion. Since you started reading this sentence the Earth has travelled 100 miles around the Sun; the Sun has moved 1,000 miles in its circuit of the galaxy; and the Orion Nebula has distanced itself another 100,000 miles from us. The Universe is restless. Everything everywhere is on the move. But not everything within the galaxy is moving away from us. Not only do asteroids cross our path locally, but on a grander scale the constellation Hercules is currently moving towards us, getting 15 miles nearer each second.

The galaxy is not a uniform mix of stars, each neatly away from its nearest neighbours. Around our neighbourhood things are rather quiet. Within 17 light years of the Sun only 45 stars are known and there is no likelihood of one of them disturbing us in the immediate future. But not far away things are different. There are clusters, such as the Hyades and the Pleiades, both visible to the naked eye, in which up to 100 stars are in close association. These two clusters are a little over 100 light years away.

Although the stars near us appear to be randomly distributed, at larger distances there are many clusters like these. The positions of these clusters are *not* random. They all lie in the Milky Way, the flat plate of our galaxy. This flat plate is called the galactic plane and forms the central line of the Milky Way, which we see as a bright arc across the night. This plane is indeed very flat. It is 500 times thinner than its diameter; it is 15 times more narrowly confined than are the planets in our solar system. The members of the clusters are very young, most having shone for less time than humans have been around. They are so loosely bound to each other that one escapes every 100,000 years. At this rate of evaporation the clusters cannot survive more than 100 million years. So clusters of stars must be forming and evaporating continuously.

As well as these young clusters of 100 stars, there are also sparse clusters of up to 1 million stars. These spherical groups are called globular clusters. They are very old and close packed. Whereas the small clusters are over 300 light years apart and spread throughout the Milky Way plane, the globular clusters are concentrated in a sphere around the centre of the galaxy. And at the heart of the whole maelstrom is a black hole, a region where gravity is so strong that even light cannot escape its pull.

If we could look at the galaxy from outside we would see a spiral structure much like the familiar shape of the Andromeda Nebula. The dense catherine wheel type spirals are called the 'spiral arms'. These are regions of compression where the fast-moving stars come to a traffic jam, entering at the rear and eventually passing out through the front. Careful measurements of the positions of many stars and clusters show us that our galaxy is like this. Moreover, we can look at distant galaxies through our telescopes and find many others that have this appearance.

There is no doubt that over half the stars form a flat disc that we call the Milky Way, and that they are rotating about the galactic centre much as the planets orbit the Sun. But not exactly like the planets. There is an important difference that is not yet fully understood and which could have implications for the future Earth.

The planets orbit around a central Sun – the bulk of the solar system is in the centre; there is nothing competing on the periphery. One result of this is that you move slower the further out you are. Thus Mercury moves very fast, Earth moderately so, but distant Uranus moves at a much more leisurely pace. If the stars in a galaxy were orbiting a central mass they too would obey the same rule (close-in move fastest, far-out slowest). But this is not what happens in practice. The stars on the edge are moving almost as fast as those nearer the centre. In the periphery where the central gravity is feeble, a fast mover should escape, flying off the spinning wheel. Yet somehow they manage to remain attached. The galaxies appear to be almost rigid bodies, as if space is filled with unseen massive agents all contributing to the general tugging, speeding up the outer regions at the expense of the centre and gripping them permanently in the system.

What is this dark matter? You may well ask. This is currently one of the forefront questions in astronomy and in high energy particle physics. One possibility is that there is a whole new species of matter pervading the Universe that we have so far been unaware of. This will be described in Chapter 11. A more down to earth possibility, if that is the right metaphor, is that there are a lot of objects made of ordinary matter, but cold and dark, not hot enough to shine. These could form balls of gas like Jupiter, but much larger.

If there are such stars, then how many are there around here? Some astrophysicists suspect that the Sun has a dark companion, Nemesis. This could well be the nearest star to us, less than 100 light days away. In cosmic terms that is in our own backyard.

The solar system and Nemesis, if it exists, travel around the galaxy together. If they encroach on other stars, their motions will be disturbed. In turn Nemesis could disturb the Oort cloud of comets and tip millions into the solar system.

So in the long term we need to know what the orbit around the galaxy is like, what hazards we can meet along the way. The annual orbit of the Earth around the Sun in one year takes us through rings of small debris which give rise to meteor showers. The journey around the centre of the galaxy takes 200 million years and much more exotic possibilities arise. Recorded history spans 10,000 years, a mere one ten-thousandth of an orbit. So life has witnessed but a few hours on a summer's day and we have no experience of the depths of winter.

During this voyage we pass through empty regions and also

several densely packed regions. We encounter one of the 'spiral arms' every 60 million years or so, and the chances of a collision are greater at these times.

At present we are in a quiet region where the risks are very small. What does the future hold? What clues do we have as to the galactic December? Clues come from the past. The Earth is 20 galactic years old and so has passed through the spiral arms on many occasions. Although humans were not here to record it, fossils provide clues of what things were like.

Dust to Dust

Every scientist, or indeed anyone fascinated with Nature, can recall the magic moment in their childhood when their wonder began. Jean Heidmann, astronomer at the Paris Observatory, recalls how one evening, just after sunset, red Mars and Venus brilliant white were close together in the twilight. His father said, 'Three months ago we were up there, between Venus and Mars'. It was an astonishing revelation to the youngster that he had travelled so far in such a short life. Spaceship Earth had transported him, along with the rest of us, through three-dimensional space and now he could look back on where we had been.

There is a vast amount of 'space' and the planets, stars and galaxies are an insignificant fraction of the total volume. The space between these three is referred to as 'interplanetary', 'interstellar' and 'intergalactic' respectively. But none of it is empty. We have already met the solar wind of subatomic particles that permeates the solar system and forms most of the interplanetary medium. The interstellar medium consists of gases, mainly hydrogen and helium and minute dust motes.

Although space is not empty, the density of gas is in most places exceedingly thin, much rarer than can be produced in the best vacuums on Earth. If you shone a light beam out across the galaxy it would encounter more gas on its way out of the Earth's atmosphere than in the rest of its journey across the galaxy and out to the vastness of the Universe.

The volume of the galaxy is so huge that although the gases are very thin on the average there are vast amounts in total. Add it all together and it amounts to more than 10 billion Suns in mass, about 10 per cent of the total mass of the entire galaxy. Most of it is in the spiral arms of the galaxy, in a layer no more than about 100 light years across. Some clouds of gas are so opaque that they prevent light from distant stars getting through – the 'coal sack' in the Southern Cross is an example. They contain the same sorts of gases as bright clouds, or 'nebulae', but these latter have stars nearby that illuminate them.

As the Sun and its attendant planets voyage around the galaxy they periodically encounter dust clouds and pass through them. We are doing so at the moment, though it is only a thin cloud and hasn't noticeably affected us. Indeed, it is providing a unique opportunity to learn about the way that the Earth and Sun would be affected when we pass through a dense cloud, such as one of the galaxy's spiral arms.

The cloud that we are currently passing through is nothing like that. There is no more than one atom of hydrogen or helium in every 10cc volume. We are travelling through at 12 miles per second about 50,000 mph, which is about two-thirds of the speed that we are orbiting the Sun. The cloud is blowing from the direction of the constellation Centaurus, heading towards Cassiopaeia. It is as a result of this discovery that we have, in only the last 15 years, begun to understand the interaction between the Sun and the interstellar medium.

If a star that is five times hotter than the Sun is encompassed by such a cloud, its heat radiation can strip the hydrogen atoms of the cloud, leaving a gas of free negatively charged electrons and positively charged protons known as 'plasma'. This zone of plasma is known as the Stromgren sphere and can reach out 100 light years, encompassing several stars. The Orion Nebula is an example where hot young stars and the interstellar medium coexist. The resulting light show makes the Orion Nebula one of the splendid sights of the January sky, and is a favourite pin-up in posters and astronomical picture books.

Rockets have taken sensitive detectors high above our atmosphere and recorded the effect of the dust cloud on sunlight. Its behaviour is nothing like the Orion glow. Instead the instruments recorded an intense diffuse ultra-violet light with a wavelength characteristic of excitation of neutral unionised hydrogen. This showed that neutral hydrogen exists in considerable amounts in interplanetary space.

At first this was a surprise because scientists had believed that any hydrogen in the vicinity should be ionised by the Sun. Then Hans Fahr and Peter Blum at Bonn University came up with the solution. Interstellar gas engulfs the solar system, but we are rushing through it so fast that the gale blows far into the solar system before being ionised by the Sun's rays.

Whereas a meeting with a tenuous cloud is hardly noticed, an encounter with a dense cloud, as in the spiral arms of the galaxy, can lead to global change.

Imagine that you are sunbathing on the sort of summer day that is familiar in Britain. The temperature is about 20 degrees Centigrade and then a cloud crosses the Sun. The temperature falls, just a little but you are already calculating the speed of the clouds – 'should I lie here and sweat it out' (wrong metaphor) or 'do I get up and put a shirt on?'

Well, that is what just happened to me and, as I write, the cloud has passed and it's sunny again. There are more clouds still to come and I wonder what it would be like if clouds were around permanently.

Sometimes in Britain or Northern Europe it feels as if that is how things are all the time! But one can always go to southern France, California or the equatorial regions and get the warmth. Suppose that there were clouds between the whole Earth and the Sun continuously. There would be a global cooling of a few degrees. You are sensitive to a cooling of just a couple of degrees. This is the subtle difference between the swimming pool being tolerable and numbing; between sunbathing and going inside; between comfort or putting the heating on indoors.

The clouds above my head are a few hundred metres high at most. The golden sun is 100 million miles away, hundreds of millions of times more remote. If that intervening space were filled with dust clouds then we would be in for a significant cold snap.

That is what happens every 100 million years or so when we encounter the dense spiral arms of the galaxy.

A spiral arm is a region where short-lived bright stars are illuminating large clouds of interstellar dust and gas. Stars and gas must be moving through the dense spirals because otherwise the spirals would become tightly coiled from the differential rotation rate of the galaxy as one moves further from the centre. The fact that the spirals survive implies that they are some wave pattern moving through the disc of stars and gas, like a traffic jam on a freeway. Traffic arrives at the rear of the jam while the vehicles at the front escape. Viewed from above over a period of time you see the traffic jam move backwards until the once-captured car finds itself at the front and escapes to freedom. The traffic jam, and the spiral arms, are patterns through which the objects move.

From observation of the spiral structures astronomers have found that objects enter an arm at its inner or concave edge where there is a dust build-up. This is a shock wave effect whereby the clouds are compressed for a short while as they cross this dust lane. The compressed clouds are almost certainly the sites of new star formation. The brightest of these stars illuminate the arm as they move along their way, but burn out before emerging into the region between arms.

The material orbiting around the galaxy may pass through these compression waves many times before being squeezed enough to collapse into a star. At each passage through an arm only a small percentage of the diffuse material forms stars or else the arms would disappear.

Our Sun and the planets were formed in a compressed cloud about 5 billion years ago. Since that time we have all circled the galaxy 20 times and encountered spiral arms on some 50 occasions, that is roughly once every 100 million years. It takes 10 million years to cross the main part of an arm and for 1 million of those we are in the compression lane.

Several people have drawn a connection between the great ice ages and encounters with the dust clouds in the spiral arms.

The ice epochs occur roughly every 250 million years and last for a few million years. They contain several glaciations of about

100,000–200,000 years each. The most recent glaciation ended some 11,000 years ago.

These facts all fit in with an encounter with a dust cloud at a spiral arm. We are currently at the inner edge of the Orion arm, having entered a compression lane 1 million years ago and recently exited. And we have only recently emerged from a period of glaciation. The time it takes to pass through the clouds also fits in with what we know about glaciations. The sun is moving at about 12 miles per second relative to the nearest stars and so at 3–15 miles per second relative to individual dust clouds. There are lots of tenuous clouds (such as the one that we are passing through at the moment) which do not affect us, but a dense cloud, typically 1 light year in size, is another matter and, at a speed of 12 miles per second, it takes us 50,000 years to get through.

Fred Hoyle and R. Lyttleton argued as long ago as 1939 that there was a connection between the ice ages and such galactic encounters. William McCrae has recently developed and extended these ideas and argues that there is a strong case for believing that ice ages and encounters with interstellar dust clouds are connected. The way in which the Sun's radiation triggers the ice age is still conjectural; the meteorology is still poorly understood. But there are compulsive features. Very dense clouds DO exist and the Sun must go through them. When it does, its radiation must be affected. The periodicity and extent of the glaciations fit in with these encounters too.

If the great ice ages are explained this way then we don't have to appeal to more sensational ideas such as that the Sun is intrinsically variable. We need not anticipate a major glaciation for a long time yet.

Encounters with the spiral arms may also be the cause of comets raining into the solar system. We are in a quiescent epoch at present, but when we encounter a dense cloud next, we may expect an increase in the risk of collisions.

There is some evidence that extinctions of flora and fauna have occurred at roughly 150 million year intervals since the Palaeozoic era 600 million years ago. Lunar rock samples suggest that meteor collisions with the Moon peak on a similar time-scale. The buffetings from the last such passage occurred some 60–70 million years ago when animal life had just emerged. At this epoch the dinosaurs disappeared. If this is all true then we are midway between such transits. The million years since humans have lived here has been a quiet period. We have tended to generalise and assume that this is the natural order of things. It will not always be so peaceful.

Lifestyles of the Stars

Nature operates on a whole range of time-scales. Stars live for billions of years, humans for decades, insects for only a few hours. But living on one scale need not prevent you being aware of the evolution of others. Let's come down to Earth and imagine a cosy peaceful scene.

It is a warm summer's day. A family is picnicking at the side of a river. Baby sleeps in the afternoon sunshine. Older children are playing with their parents. Grandparents are dozing. Hovering above the lilies, dragonflies enjoy their brief moment of existence. Their lifespan is but a millionth part of a human life. An average human life is but a millionth part of a geological epoch, which is in turn 1 per cent of the age of the Universe. So as the dragonflies are to the humans on the riverbank, so are the humans to the stars.

Suppose that dragonflies were highly intelligent. In their moments of existence they would be aware of the seven ages of man at the picnic party. Although their own life is very short by comparison, they would see the evidence of much longer time-spans. They will see the development that individual humans have already experienced, or have yet to go through.

We human insects in our turn perceive the grander epochs of time. Our three-score years and ten are like a dragonfly's afternoon. When we look at the stars we see the past and future of our Sun. There are misty nebulae, stars in gestation; middle-aged ones like the Sun today; old ones, the future dying Sun.

For all stars are not alike – you can see this with your own eyes. Orion is one of the best-known constellations. If you live in the northern hemisphere, then on a clear winter's night you will be able to find bright red Betelgeuse at the top left corner of the shoulder. Nearby is Sirius, a bright blue-white colour. Vega in Lyra can be seen blue in the summer sky. In the southern hemisphere you can see the Southern Cross. This contains three bluish-white stars and one red star.

With binoculars you can see millions of stars in a host of colours. The colours tell you about the temperature at the surface of the star. Just as an electric fire glows red-orange, then yellow as it warms, so the hotter the stars, the further across the spectrum the colouring is. Blue Vega shines at 30,000 degrees, our yellow sun is some 6,000 degrees, while red Betelgeuse shines at 3,000 degrees.

You can feel the warmth of a fire before it begins to glow. The same is true of the stars. Stars that are too cool to emit visible light none the less emit heat rays, infra-red radiation. In 1984, IRAS surveyed the skies and sent back vast quantities of data about fledgling stars of the future. The astronomers are still processing this wealth of new information through their computers but already they have learned a lot of things about star formation.

How are stars formed?

No one has yet given a complete detailed and totally accepted answer to this question. However, from watching the behaviour of many different types of star in the distant parts of the cosmos, and putting together the various clues and experiences gained from decades of observation, we have a pretty good idea of the general scheme of things.

Nature's cycle is of life leading to death, which in turn seeds new life. The seasons on Earth are a rapid and small-scale cycle compared to the grander and slower scale in the Universe at large. The deaths of some stars give birth to the next generation. Catastrophic supernova explosions (see Chapter 9) can destroy nearby planets and life (if there are any elsewhere), just as a conventional bomb blast can destroy solid matter in its vicinity. These cosmic blasts also send out shock waves through the interstellar gases, compressing them. In some regions the gas becomes concentrated enough that clumps form. The patchy mist thickens here and there.

Gravity begins to exert its influence, tugging these clumps feebly but insistently towards one another. Gradually, over thousands and millions of years the gas collapses into a huge ball. Its own weight continues to crunch it smaller and smaller. This would go on until no room is left except that something happens to prevent it. The random jostlings of the atoms generate heat and light – the gas begins to shine. The heat within resists gravity's inward pull.

The temperature of the gas in its quiescent state was 3 degrees above absolute zero; a chilly −270 degrees Centigrade. As the atoms begin to bang together, the temperature rises, eventually reaching temperatures familiar on Earth. These are still far too low to glow visibly, but already gas is emitting low-level heat radiation, infra-red radiation.

Very light clouds of hydrogen gas don't have enough weight to collapse under gravity's pull and so they never get hot enough in the middle for thermonuclear fusion to begin. These failed stars are clouds of cold gas like Jupiter. There could be a lot of these 'dark suns' around, such as Nemesis, a suspected dark companion of our Sun. These dark stars have been invisible in the past; science is only now beginning to be able to detect them by their infra-red heat.

Special heat sensitive cameras can 'photograph' objects by the heat they give out. They can image the human body by its heat. Infra-red cameras have been sent up on satellites and sent back images of gases beginning to warm up in this way. At this stage we say that it is a 'protostar'. Finally the gas is warm enough that the infra-red radiation gives way to a dimly glowing visible red.

But gravity still beats the feeble glowworm, squeezing the star until nuclear reactions begin. Now at last a fully fledged 'main sequence' star erupts. Nuclear reactions take over. Hydrogen nuclei fuse to make helium. The temperature rises to a staggering 10 million degrees or more. These are bright spots that we call stars.

So stars are in a state of internal conflict: gravity pulls them in while thermonuclear fusion keeps them alive. The outcome depends on the amount of gravitational pull (the size) and the status of the nuclear reactor (how much useful fuel is left and of what type).

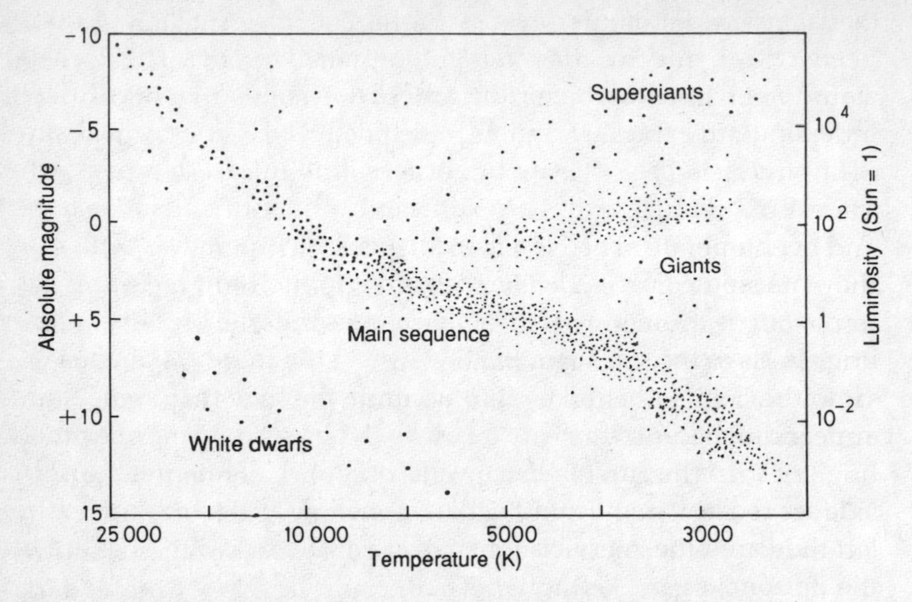

8.1 Hertzsprung–Russell diagram of star types The absolute magnitude or brightness of the star is plotted along the vertical axis and its temperature along the horizontal. Brightest are at the bottom and hottest to the left. Cool stars on the right shine red; hotter ones to the centre are yellow; and very hot at the left are blue. Stars are not randomly distributed. Many are grouped in a narrow band on the diagonal called the main sequence. Another group, known as the giant branch, extends horizontally above this. The dim cool stars are very big and are called giants and supergiants ('red' giants because of their colour). The very bright hot small stars are the (blue and white) dwarfs. Our Sun is presently near the middle of the main sequence.

It has taken only a few million years for the interstellar cloud to condense and start to shine as a star. I say 'only' because it is short in cosmic time. For example, this is the sort of time that humans have existed on Earth. Interstellar clouds that condensed when Neanderthal man walked the planet are now shining in the heavens.

Hans Bethe, a German-born theorist at Cornell University in the USA, worked out the sequence of nuclear reactions that drive the stars. This explains the varied stars that we see. It has made us realise that the stars in the sky are not permanent, but are continuously changing. In particular, it enables us to work out the future of our Sun.

To place our Sun's present position in the stellar evolution needs a scheme for classifying the stars we see. The Danish astronomer Ejnar Hertzsprung and the American Henry Russell independently (in 1911 and 1913 respectively) came up with the idea of classifying stars by brightness and surface temperature. These two characteristics are sufficient to identify a star's status and eventual fate. The Hertzsprung–Russell diagram is a chart of the stars (see Figure 8.1). The scale from left to right measures the temperature; from top to bottom measures the true brightness (the brightness of the star were we to view it at the same distance as we view the Sun – this takes into account the fact that some stars appear dim because they are far away, whereas they may be much brighter than the Sun when viewed close to). Thus stars at the right side are red hot; in the middle they are yellow and white hot; to the left they are blue. Very bright stars are at the top; dim ones are at the bottom.

The Sun's brightness puts it about halfway up the chart, and its heat puts it about halfway along the chart. You do the same for each and every star and you find them scattered all over the plot.

Well, not quite all over. What you immediately notice is that the stars don't occur at random. The majority lie on a line, known as the main sequence, which includes the Sun. A handful of dim white stars, the white dwarfs, is at bottom left. Rather more red giants are at the top right corner.

The amount of time a star spends in the various regions depends on how big it is. A massive star will be sucked tighter under its own gravity than a small one. Its collapse is prevented by the heat generated from burning its nuclear fuel. As it burns, it gets hotter to remain stable. The hottest brightest main sequence stars are the heaviest (top left-hand corner). They are burning fuel so quickly in generating the heat that their lives are the shortest. Their deaths can be dramatic and the sequel bizarre, as we shall see in Chapter 9.

Less bulky stars burn their fuel more sparingly: their gravity is less crushing, less heat is required to hold them up, and they live longer. Our Sun is such a star.

A star like the Sun can shine like this and be quite stable for 10

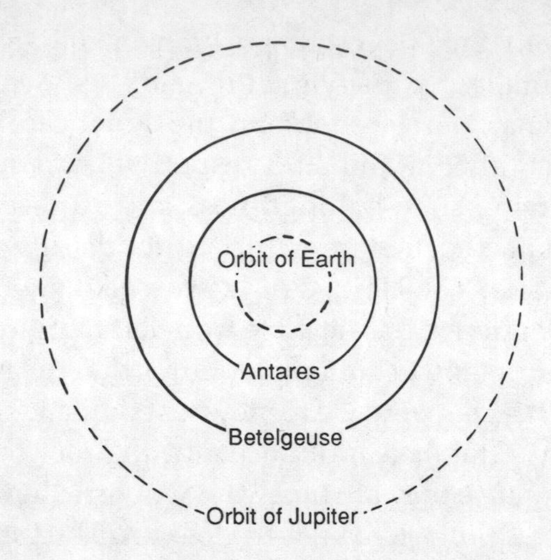

8.2 Sizes of stars There is an immense variation in the sizes of stars. Those with diameters of about half a million miles, like our own Sun, are quite common. The largest stars are red giants and supergiants. Antares has a diameter 300 times that of the Sun and if placed at the centre of the solar system it would swallow the orbits of Mercury, Venus and the Earth. Betelgeuse in Orion is even bigger. At the other extreme, white dwarfs are about the size of the Earth, while neutron stars are only a few miles across.

billion years. It converts 600 million tonnes of hydrogen fuel into helium every second, radiating energy in the process. There is enough hydrogen for this to continue for another 5 billion years. The solar system is about 4½ billion years old, so the Sun is about halfway through its hydrogen stock.

When it has used up all of the hydrogen in its central core, the thermonuclear reactions will spread outwards. In the process, the Sun will expand rapidly a hundredfold and become a bright red giant. The inner planets wil be engulfed; vaporised. The Earth will cease to exist.

In the red giant phase it will be unsteady for some thousands of years. It may become a 'variable' star, expanding and contracting every few hours. The pressure of the gas within the star pushes it outwards beyond its average size until gravity pulls it back. It keeps overshooting and falling back like a swinging pendulum. There are many examples of such stars in the sky. Betelgeuse, the bright red star in Orion, varies gradually in size from 300 to 400 times the diameter of the Sun. (To gain some feeling for what this means, it is as if the Sun consumed the Earth, extending beyond the orbit of Mars.)

Finally all nuclear burning in the core will end. No fuel remains to resist the gravitational force. The star collapses under its own weight. It becomes a white dwarf. There is no source of energy within to generate new heat and light, but even so it cools so slowly that it shines for another billion years.

So there will come a day, in some 4 billion years, when the Sun will rise for the last time over the eastern horizon. When an audience in Washington learned that the Sun had 4 billion years left, one member asked, in a slight panic, 'Did you say 4 billion or 4 million?' On being assured that it was 4 billion they sat down, at ease.

This shows how difficult it is to gain a feel for large numbers, so let's describe the Sun's lifestyle on the more familiar time-scale of a human span: a century.

Its gestation, the protostar phase, lasts a mere two days. It spends 80 years on the main sequence, consuming its hydrogen fuel. When this is through it becomes an unsteady variable red giant for a fortnight and then spends eight years in retirement, living off its reserves until its ultimate death. On this time-scale the Sun is now about 40 years old – halfway through its present 'active' phase, at which time intelligence arrived on Earth: life begins at 40!

The Sun is typical of many stars; that is why we can make such confident assertions about its future life cycle. It gives us ease to realise that the Sun has a healthy long-term future, and that the Earth will not suddenly turn into a stellar furnace of its own accord (another Hollywood myth thankfully disproved).

However, some stars in the sky will change form catastrophically. If this happens to a star in our vicinity the consequences could be very grave. Although our Sun will form a red giant and then collapse into a white dwarf, it will do so only slowly. If the Sun had been 50 per cent heavier, it would have gone through these stages much faster. Indeed it might already have died, its collapse having run out of control and a nuclear blast shot out across space. The end result of such a catastrophe is a neutron star – an atomic nucleus the size of a city. If the starting star was even heavier, then it may end up as a black hole in space.

So let's learn more about the neutron star and the supernova blast that spawns it.

9
Exploding Stars

In 1967 Jocelyn Bell was a student of astronomy at Cambridge where she was working under the guidance of Antony Hewish, one of the leading radio astronomers of the day. Rather than observing stars by optical means, they were looking at the sources of radio waves coming from the cosmos.

Day by day the radio telescope peered at different parts of the sky, into the depths of space. It was fairly routine stuff. Their idea was to measure the size of the sources, 'quasars', by seeing how much they twinkled. You are familiar with the fact that stars twinkle as their light passes through the atmosphere, whereas the nearby planets, with their larger angular size, do not twinkle. By analogy, the radio waves from relatively small sources will twinkle as they pass through the solar wind whereas those from larger sources will be less disturbed.

Bell recorded lots of twinkling radio signals until one day she noticed one that looked odd. It seemed 'scruffy', for that is how she described it to Antony Hewish as she discussed with him what to do about it. He suggested that she set the apparatus up to be able to record the twinkling at a faster rate – to resolve the radio signal as it flickered back and forth. When she did so she discovered that the 'scruff' resolved into a series of pulses, regularly once a second or so.

They thought that they had picked up the signal from some radio beacon but after detailed enquiries it became clear that this was caused by no terrestrial source. What could be going on?

Members of their department jokingly suggested that they

were signals from LGM – little green men – which the media seized upon and propagated. Hewish said that if the signal was real then it was most unlikely that they had stumbled across the only one in the entire Universe and suggested that she search for more. Soon a couple more examples turned up and the 'pulsars' became official.

They had discovered a special type of star which beams out its signal like a lighthouse, whirling the light round and round. You see a flash only when the beam points at you. It circles around, away from you, and you see nothing until it points at you again.

Bell's original sightings were of stars that emit radio waves, and so had been picked up by the radio telescope. Today we know of pulsars that emit visible light and flash on and off, up to 30 times a second as the beam whips round and round. The culprits are a special kind of star that consists entirely of subatomic particles called neutrons. Their existence had been suspected for 30 years but no one had anticipated the pulsing lighthouse signal. Hence the initial confusion when Jocelyn Bell discovered the first one.

Neutron stars are the dense cinders of heavier suns, the end-products of a catastrophic explosion which destroyed a huge star. The outer mantle of the star erupts into space, covering an incredible 30,000 miles in the first seconds. The remaining core is a dense pack of neutrons – the mass of the sun compressed into a ball no bigger than New York.

Neutron stars are fascinating objects. The blast that spawns them can light up the sky: the brilliant sight is called a 'supernova'. (The Latin for new is 'nova'. The flash from the white dwarf is superbly bright; hence the name 'supernova'.) It is like witnessing a gargantuan nuclear blast across space. If one happened in our vicinity it would mean the end of life on Earth. If the nuclear radiation didn't finish us immediately, the shock wave and debris following on behind would disrupt the atmosphere.

On 23 February 1987 a supernova burst out in the Large Magellanic Cloud, a nearby satellite galaxy to our own. This is 200,000 light years away; far enough to be no harm while near enough to excite astronomers and physicists. In the southern hemisphere it was visible to the naked eye – the first such since the invention of the telescope. This highlights how important it was;

everyone has been watching it closely. More about the sighting and what we have learned later; first – what do we know so far?

Route to the Neutron Star

Have you ever been at the bottom of a pile of people, underneath a collapsed rugby scrum for example? As more and more weight piles on top, the pressure becomes unbearable. Think what it must be like at the base of a mountain. The pressure of the overlying rocks, piled miles high, can be literally crushing.

The summit of Mount Everest is 5½ miles above sea level. The island of Hawaii is the tallest mountain from its undersea base to the top, some 6 miles. On Earth no mountain could survive more than 12 miles high. The material at the base would flow like liquid owing to the pressure of the rocks on top.

Gravity is an inexorable crusher. Thermonuclear reactions in stars resist its force but even so the centre of the Sun is 100 times denser than water. When the nuclear fuel is exhausted, stars can no longer support themselves against gravity. In extreme cases, such as neutron stars, a thimbleful of matter can weigh millions of tonnes.

Matter at extreme densities takes on forms that are unfamiliar on Earth.

There are many different elements that go into building up the world around us. But they all have one thing in common. The matter that we are made of is remarkably empty. You can compress mud into small balls with the strength of your hands. Even metal can be compressed under the pressure of pile drivers. The rocks under mountains may be liquefied by the pressure. At the atomic level there is more empty space than matter! Less than a billionth part of the atomic volume contains matter – electrons encircling a nucleus of protons and neutrons.

To get an idea of how empty the atom is, imagine the diameter of a hydrogen atom (one electron orbiting a single proton) scaled up to the size of a 500 metre long hole at a golf course. This is the longest fairway that you will find and a top-class golfer will take three or four good shots to reach the green from the tee. The goal is a tiny hole, some 2 or 3 cm at most, known as the pin. The size of the pin relative to the length of the huge fairway is as the size of the nucleus relative to the extent of that atom. All of the distance from the tee to the green and to the pin is as empty space in the atom. (If you are a football fan, you can imagine a pea at the centre spot of a pitch.)

An atom is about one-hundreth of a millionth of a centimetre across. This is unimaginably small. Yet its nucleus is 100,000 times smaller. Whereas electric and magnetic forces bind the electrons loosely in atoms, the nucleus is gripped by powerful forces extending only one millionth of a millionth of a millimetre. At these short distances the nuclear forces dominate over everything else – gravity and electromagnetic forces are trifling by comparison.

An atom's size is determined by the lightweight cloud of electrons whirling in the peripheral regions. These govern the chemical behaviour and physical properties of matter on Earth. In our daily experience we are aware of electromagnetic forces and the electrons, carriers of electricity, but the nuclear forces do not directly affect us (though the waste products of radioactivity cause immense political problems). The protons carry positive electricity, which balances the negatively charged electrons in neutral atoms. We can think of the protons as ensuring the neutrality of atoms, and the neutral neutrons being needed to stabilise the nucleus. Together they provide 99.95 per cent of the mass of bulk matter, such as you and me.

So we are made of atoms and our mass is concentrated in less than a billionth part of their volume. Put another way, the density of nuclear matter is a billion times more than the matter that we are familiar with on Earth.

We can squeeze atoms nearer to each other but cannot compress individual atoms. Their sizes are fixed by Nature and depend on unchangeable constants such as the strength of the electromagnetic forces and the mass of the electron.

But gravity is the ultimate compressor. As we add more and more material, the pressure of its weight becomes so intense that the atoms fragment. Their electrons no longer remain in their orbits but are displaced. Instead of matter containing atoms built of nuclei and orbiting planetary electrons, we have nuclei sitting in the midst of a dense homogeneous gas of ubiquitous swarming electrons. This form of matter, plasma, is the most common form of matter in the Universe. It is we on Earth, with the beauty of diamonds and crystals, with chemistry, biology and life, who are the exception. Plasma rules.

This gas of free electrons is what would occur if we kept on adding weight to the Earth. The atomic electrons would not be able to sustain the ordered structures. The solid Earth would give way to a dense homogeneous gas of electrons and nuclei.

As we learned in Chapter 2, debris from space is continually falling to Earth, adding to the Earth's mass. However, this is such a trifling amount that we needn't worry. We would need to exceed the mass of Jupiter by a lot before the solid Earth changed its state.

Once the atoms have been disrupted they can no longer resist the crush. If more and more matter were added the result would be a white dwarf. This is a star as massive as the Sun but as small as the Earth. There are no free protons available in the gas of electrons and atomic nuclei to fuse and sustain thermonuclear reactions. One tonne is crammed into each cubic centimetre (less than a teaspoonful).

What happens if we go further? The atomic volume, once disrupted, can be compressed 100 billion times. The pressure that gravity brings to bear is 1 million billion times greater than anything ever achieved in an Earthbound laboratory, so we need not fear a self-induced Armageddon by some mad scientist crushing matter and changing the nature of the Earth. We cannot yet change the stars nor planets.

We cannot, but Nature can and does.

In white dwarfs the density is 1 million times that of water. At first no one understood how these extreme densities could be maintained without the star collapsing. Then, with the discovery of atomic structure and the rules governing the stability of atoms, the stability of white dwarfs was in turn explained.

The electrons are moving relatively slowly in this gas. As the density increases, the electrons speed up. Soon they are moving near to the speed of light and the mathematics no longer applies. Instead of using the laws of low speeds we have to go over to high speeds, the province of relativity. This requires a few changes in the mathematics and, in 1930, Subrahmanyan Chandrasekhar worked out the consequences.

He did this in an unusual set of circumstances.

Chandrasekhar, still only 19 years old, was on a boat en route from Madras to England where he was going to study. To pass the time he calculated what would be the effect of increasing the density in a white dwarf. He found an astonishing result: white dwarfs cannot exist if their mass exceeds that of the Sun by more than 40 per cent. The effect of relativity is to enfeeble the electrons' resistance to the inward pull of gravity. If the mass gets big enough, bigger than the 'Chandrasekhar limit', gravity wins out and the star collapses; the white dwarf can no longer exist.

Chandrasekhar checked his calculations and couldn't find any mistakes. Next he began to wonder what would happen to a would-be white dwarf that had too much mass, too much inward pull to survive. He thought that it would collapse and become a black hole, a dense star where the force of gravity is so powerful that it drags back even light before it can escape.

As the boat docked, Chandrasekhar possessed a secret of the Universe that no other living soul knew. He told Arthur Eddington, one of the leading physicists of the time. Eddington disbelieved it. So did other senior scientists. Who was this 19 year old calculating on a boat trip and announcing the death of stars? It was all a bit rich.

But Chandrasekhar was right, as people gradually recognised. It is all very well with hindsight to criticise Eddington and others, but the claim was quite outside current thinking, star shattering indeed. Many years later, Chandrasekhar's genius was recognised; for this work he won a share of the Nobel Prize in 1983.

Chandrasekhar was wrong on only one thing – his belief that the overweight white dwarf would collapse into a black hole. This wasn't a mistake; at the time it was the logical conclusion, and may in part have been the psychological barrier that prevented scientists immediately accepting his theory. No force was then known that could prevent the electrons and protons in the star collapsing inwards under gravity's pull. In 1931 electrons and protons were the only known atomic particles. No one yet knew that a neutron, an electrically neutral version of the proton, exists in atomic nuclei. The neutron was discovered only in 1932 and its existence provided the crucial missing ingredient in the story of the stars.

Soon after Ernest Rutherford identified the existence of the atomic nucleus in 1911 and of positively charged protons, he had suggested that there could exist a neutral particle formed by a negatively charged electron being absorbed in a proton. Several people then speculated that there are two varieties of nuclear particles – the positively charged proton and a neutral counterpart, the neutron. Marie Curie's daughter, Irene, had almost discovered it early in 1932 but had misinterpreted her results. Rutherford, at that time a professor in Cambridge, had stimulated his colleagues into being on the look out for such clues and his deputy, James Chadwick, immediately realised that Irene Curie probably had been producing neutrons. He set to work at once and almost overnight produced, identified and proved the existence of neutrons.

Telegrams announcing the discovery went to the leading scientific establishments around the world. In Copenhagen Lev Landau, a leading Soviet theoretician who was visiting at the time, immediately started to consider the implications of the discovery. That same day he gave a seminar and announced that neutron stars should exist.

Here was the missing link in Chandrasekhar's theory. When the overweight white dwarf undergoes a catastrophic collapse, its electrons and protons encroach on one another to the point that they fuse into neutrons, and the collapse stops. The remnant of the 'supernova' explosion is the tiny neutron star.

So a white dwarf that is slightly heavier than Chandrasekhar's limit (40 per cent heavier than the Sun) will collapse and end up as a neutron star, no more than a few kilometres across. If the white dwarf is less bulky than this, say the same sort of mass as the Sun, then it may remain a white dwarf or end up as a neutron star (which of the two depends on what other disturbances it experiences). Typically the size of a solar-mass neutron star will be 12 miles diameter – the mass of the Sun contained in a sphere whose size is like Amsterdam or New York. If the mass exceeds 3.4 solar masses, then even the neutrons cannot prevent the collapse and, barring new unknowns, the neutron star *will* fall into a black hole.

If you were on the surface of a neutron star, the force of gravity would be 100 billion times greater than you experience on the Earth. Whereas on the moon the gravity is less and you can jump higher (anyone could break the 'world' high jump record on the moon!), on a neutron star the gravity is so intense that your head would weigh as much as 100 ocean liners. Indeed, mountains could not resist the downwards pull: even a Mount Everest would be under 1 metre high. Its head would stick up through the atmosphere, which would be all of 5 cm thick. Mountaineering would be an exhausting enterprise as it would take the energy expenditure of a lifetime to climb 1 centimetre.

This is what the astrophysicists claim, but how sure can we be that they are right?

In the 1930s, neutron stars were no more than an idea. No one had seen one and most people doubted that anyone ever would, even supposing that they existed! For 33 years nothing happened, until the day when Jocelyn Bell noticed a regular bleep coming from a distant star as if something, or someone, was sending a signal.

Neutron stars exist. The remains of supernovae, the evidence of past cataclysms, are out there. From their numbers and the age of the galaxy we deduce that a supernova occurs somewhere on average once every 20 years. What are the chances that one will occur near enough to be a hazard?

Supernova

What were you doing at 07.30 GMT on 23 February 1987? I was having breakfast as, unknown to me, a burst of neutrinos passed through my cornflakes. All the time we are bathing in a breeze of neutrinos coming from the Sun; but the sudden burst that February morning was quite different. It was a blast from a dying star, 170,000 light years away in the Large Magellanic Cloud galaxy.

For over 25 years astrophysicists have believed that the gravitational collapse thought to be associated with supernovae and neutron star or black hole formation is a copious source of neutrinos. In fact they argued that the brilliant flash of light – the traditional manifestation of a supernova – is only a minor part of the drama, perhaps less than 1 per cent of the whole energy output. The bulk of the energy radiated by the collapse comes off in invisible form – the neutrinos.

'Invisible' in the past, but not now that we have built neutrino telescopes. The exciting news is that in this case, for the first time, we have detected neutrinos emanating from outside our galaxy (previously we had only seen those from the Sun) and have proved that the theory is right: when stars collapse they throw off their energy as neutrinos.

The theorists are still poring over the data, learning more about supernovae and the death of stars than ever before. This is the culmination of a long story that began with the idea that supernovae are the birthpangs of neutron stars.

The first hints came with a suggestion from W. Baade and F. Zwicky, German and Swiss astronomers working at the Mount Wilson Observatory in the United States in 1934. They realised that the process of collapsing into a neutron star will emit vast amounts of energy. This radiation appears as a sudden flash visible in the sky like a new star.

Occasionally such flashes do appear; In AD 1006 'a new star of unusual size', which 'dazzled the eyes, glittering and causing great alarm,' shone for three months according to reports from central Europe. This is the only example reported outside the Far East before the Renaissance, so it must have been bright. The Chinese and Arabs saw it on 30 April 1006. Could this be an example of a 'supernova'? To answer this we first need to know how much energy is emitted in a supernova.

You start off with a white dwarf – the mass of the Sun in the volume of the Earth, a radius of some thousands of miles. After the collapse the mass is concentrated within a sphere no more than about 7 miles across. The whole mass of a star has fallen through a distance of about 1,000 miles. Falling objects have a lot of energy (drop a brick on your foot if you want a demonstration!). A glass falling from a table can shatter with a crash; the energy in its motion (kinetic energy) is changed into the energy of sound waves. A meteor falling in towards Earth can burn as bright as the Moon for several seconds as energy is transformed into light. Now imagine a whole star falling inwards through hundreds of miles under the pull of gravity.

The energy emitted is enormous. We are used to hearing of megatonne (millions of tonnes) hydrogen bombs. A megatonne bomb is 70 times as powerful as that dropped on Hiroshima. The heat is 5,000 degrees, hotter than the surface of the Sun. This is a world where fire bricks melt, where steel balls vaporise. That is the result of 1 million tonnes. The masses in stellar collapse are in a different league entirely – 1 megatonne for every cubic metre in a volume as big as the whole Earth. The energy released is the same as our Sun emits in 1,000 billion years. The Sun is only about 5 billion years old at present, so the energy emitted in a supernova burst is over 100 times greater than the Sun has put out *since before the Earth began*. And the supernova does this all in a few days!

Astronomers watch the skies continuously. Suddenly, in the space of a few days they see a star become tens of billions of times brighter. For a brief period this single star can outshine an entire galaxy of stars. When a supernova occurs in a distant galaxy you first identify which galaxy it is shining from; this tells you how far away it is, as the remoteness of most galaxies is known to within a factor of two. You then compare its brightness to that of a galaxy and hence deduce the total energy that it is giving out.

We can easily illustrate that the energy output agrees with our estimates. The crucial feature is that the supernova can outshine an entire galaxy for a few weeks. A typical galaxy contains 100–1,000 billion suns. Our own Sun has been burning for about 1,000 billion days. Its lifetime output is akin to the daily quota for an entire galaxy. So if a supernova outshines a whole galaxy it is competing with the Sun in all history.

Perhaps the most famous remnant of a supernova is the Crab Nebula, which has been called the Rosetta Stone of the life and death of the stars. This erupted in 4000 BC. The blast threw off intense radiation and ejected the outer shell of the star into space. The radiation, containing lethal X-rays and gamma rays as well as visible light, set out across space. 5,000 light years away planet Earth awaited, unaware of the approaching lightwave.

Early in the morning of 4 July 1054 Chinese astronomers (actually we would call them astrologers today) saw a bright new star rising in the east just before the Sun. They called it a 'guest star'. During the next few days it brightened further until it outshone all the other stars in the sky. For nearly a month it was so bright that it shone by day as well as by night. This was the light from the supernova. After 5,000 years, travelling, the rays were too sparse to cause trouble, but the sight was awesome none the less. Gradually it faded and within 18 months was no longer visible.

Not only do we have the Chinese records of its visible light, but the Earth has preserved the record of the gamma radiation.

Go to places that are changeless and you may find ancient records remaining. The Antarctic continent is a unique example where, by digging deep down into the snowpack, scientists can examine the snowfall each year over a millennium.

Rain and snow bring dust from the upper atmosphere. The rain washes away; the snow preserves the dust. When gamma rays hit the atmosphere, they fuse the indigenous nitrogen and oxygen into nitrous oxide. The Antarctic snowpack of hundreds of years past shows an increased presence of nitrates in years that supernovae have been seen. This fits with the sudden pulse of gamma rays hitting the upper atmosphere.

The rest of the debris of the star, its outer shell, is following more slowly. The bright light is the herald of shocks to come. With modern telescopes we can today identify the wreckage from the Crab supernova. It travelled 30,000 miles in the first few seconds. It is still approaching us 900 years later, at the rate of a few miles each second, and is bigger now than the images in photographs taken in 1899. But it is still over 6,000 light years away and at this rate of progress it will not reach us for another billion years. Our descendants need not worry, as by then its force will be utterly spent. Most likely the dust will be trapped by the gravity of other stars en route or will form new stars as the shock wave perturbs the gas in interstellar space.

The dim nebulous mist was first seen in 1731 by the English astronomer John Bevis. It is in the constellation Taurus and if you have a telescope you may be able to see it for yourself. This constellation is in the same part of the sky where the Chinese saw their 1054 guest star. In the nineteenth century the English amateur astronomer, the Third Earl of Rosse, was the first to resolve crablike filaments protruding from its southern extremity. This allusion to the legs and pincers of a crab gave the nebula its popular name.

Although it is now fairly dim in visible light, it still shines brightly in ultra-violet, infra-red, X-rays and radio emissions. Indeed, it ranks with the most brilliant of all celestial objects. All in all it still shines as bright as 30,000 suns.

It is now an egg shape, 15 light years long by 10 light years across. We know that it is expanding at a few miles each second and is slowing under the tug of its own gravity. If we turn the clock back, then in the past it must have been much smaller and hotter than it is now. Calculations imply that it was a single point around AD 1000, so indeed it appears to be the remnant of the 1054 supernova.

Further proof that the theory of stellar evolution looks right came with the discovery of a neutron star at the heart of the Crab Nebula. This shows up optically, blinking on and off 30 times each second. There can be no doubt that neutron stars are the remnants of supernovae. The Crab has all of the ingredients: visual sighting of the light in 1054, the gamma ray record frozen in the snows, the debris still approaching us, and even a neutron star remnant.

After the Crab supernova in 1054 another big one occurred on 7 August 1181. The next two were very close: 1572 and 1604. Astronomy and telescopes were flourishing by then and so these outbursts were opportune and well recorded. The 1572 supernova shone as bright as Venus (brighter than anything apart from the Moon). The 1604 one was somewhat dimmer, shining as bright as Jupiter – still a fine sight. Since then we had had nothing for nearly 400 years until February 1987 when a supernova visible to the naked eye burst into prominence.

Actually the violence really took place 170,000 years ago in the Large Magellanic Cloud, a satellite galaxy of ours that is visible in the southern hemisphere. A flash of light, brighter than 1 billion suns, and a blast wave of neutrinos flew out from the debris. Travelling 10 million miles each minute they raced out from the site, left the galaxy and headed out across intergalactic space, their 1987 rendezvous still far in the future.

Ahead of them lay the large Milky Way, within which one insignificant star, the Sun, was orbited by a lump of rock on which molecules had organised into life. The most advanced of these life forms were humans who had progressed to the Stone Age.

The shell of radiation travelled onward, while on Earth humans procreated and discovered science. In the 1930s they came to realize that radioactive processes spawn neutrinos, but scientists doubted that anyone would ever be able to catch one, so feeble are the interactions between neutrinos and matter.

Meanwhile the wave from the collapsed star came inexorably on and was approaching the Earth through the southern heavens. Slightly more than 30 light years ahead of it two American scientists cleverly managed for the first time to prove that neutrinos exist by capturing a few that had been emitted by a nuclear pile. That was in 1956, since when the study of neutrinos has become routine, yet even now we don't know if they weigh anything or not.

The blast wave was some 15 light years away when Ray Davis started operating his solar neutrino detector in the Dakota mine. Although ideal for catching the breeze of solar neutrinos it is almost blind to the neutrinos coming from the supernova.

A few years ago, however, some physicists began building apparatus underground that had nothing to do with supernovae or neutrinos but which has turned out to be eminently useful as a neutrino telescope. Their hope was to find evidence of decaying protons – the whisper of a dying universe (more about this in chapter 11). To capture such a rare event it is necessary to hide far below ground where the earth above one's head acts as a blanket against the incessant bombardment of cosmic rays. Very little can penetrate to the bottom of the deep mines in which the apparatus is contained; neutrinos can though.

In these caverns scientists have constructed vast swimming pools filled with thousands of tonnes of water. If a large number of neutrinos pass through there is the chance that one or two might interact with atoms in the water and give their presence away. On 23 February 1987 the neutrinos from the exploded star, having travelled for 170,000 years, passed through the Earth onwards into space. As they did, a handful were captured in the water tanks.

The supernova has been seen shining as bright as the entire Magellanic Cloud, and is the first major supernova since the invention of the optical telescope. And for the first time in history humans have detected the explosion of a star in neutrinos. This proves that the astrophysicists were right all along about the way stars collapse, and adds to our confidence that we understand a great detail about what is going on 'out there'.

Go to the hot desert at high noon in summer and you will feel the scorching furnace of the Sun – 100 million miles away. After a few hours you will suffer from severe sunburn – the first stage of radiation damage from exposure to a distant nuclear blast. Now imagine all that power accumulating for aeons and then suddenly shot at you in an instant. If something like this occurred near the Earth, we would be utterly annihilated. The supernova was close enough to be invaluable for science, but remote enough to be no danger to us, though we will never know what other civilisations may have been destroyed within its immediate vicinity 170,000 years past.

We can survive the billion-sun supernova if the outburst is more than 50 light years away. The nearest star to the Sun is 4 light years away, though stars are on average 1 light year apart. Thankfully there are not that many stars around in the danger zone. At most one or two are likely 'future' supernovae.

Present theory implies that supernovae occur for stars that are much bulkier than the Sun. Such stars are bright; you can't miss them. You can see an example on most clear nights wherever you live. On the left shoulder of the constellation Orion is bright red Betelgeuse. This shines like a ruby and promises to be a wonderful sight should it explode; it is 650 light years away and will cause us no problem.

We do not know what other civilisations may have been destroyed, or are currently threatened by supernovae. We appear to be safe, so long as our understanding of stellar workings is correct.

Countdown

The blast on 23 February 1987 was preordained, an unstoppable sequence of events programmed millions of years before. Hydrogen burned to helium, like in our own Sun, but with ten times the mass of the Sun the core heated up to 150 million degrees, at which point helium nuclei combined to build carbon, neon and silicon. Our Sun can burn for 5 billion years more but the onset of heavy elements in the more massive star means that by this stage its death is less than a year away.

The temperature of the core is a staggering 2 billion degrees as the silicon squeezes together building up iron until the heat is like 1,000 of our suns condensed in a single star.

Iron is the most tightly bound of all atomic nuclei. It can't fuse spontaneously into heavier elements as that would consume energy instead of liberating it, and so the fusion fire goes out in the iron core. The star has less than three days left. If we could have looked inside the distant star in 1987 we would have seen the beginning of the iron core on 20 February. The silicon fusion rapidly uses up the reserves until the iron is about 50 per cent more massive again than our Sun. The star is configured like an onion skin with a dense central iron core surrounded by burning layers of silicon, oxygen, neon then carbon and finally helium and hydrogen. It no longer makes enough energy to support its enormous weight and it starts to collapse under gravity. It has less than 1/5 second to live.

A lot happens in that brief moment as the implosion begins.

The core collapses at up to one-quarter the speed of light and shrinks from a diameter of about half the Earth to only 6 miles. Its matter is shattered into its basic constituents – protons, neutrons and electrons – and is three to five times denser than an ordinary atomic nucleus. This is very unstable and the inner core bounces back, doubling or tripling in size, sending a shock wave at over 10,000 miles per second into the outer layers which are still rushing inwards. The shock wave has more energy than an entire galaxy puts out in a year.

The blast changes some of the nuclei in the outer reaches into heavy forms, leaping the iron barrier and producing elements as heavy as lead and uranium, and also some indium. It ejects them into deep space, polluting the universe with a mix of elements, perhaps the seeds of future life forms. The seeds of our atoms were formed in such blasts more than 5 billion years ago.

This has all taken place in less than 1 second. In the next 4 seconds the inner core emits a burst of neutrinos – as detected here on 23 February 1987 – as all its nuclear material turns into neutrons. It is in effect a huge atomic nucleus containing 10^{57} neutrons.

The shock wave, meanwhile, travels on outwards and exits the outer surface after about an hour. Previously hidden from view by the opaque outer surroundings, now the energy of the shock is unleashed as light. This is first ultra-violet and then the full flash of visible light that has historically been the famous sign of the supernova star.

This is now visible in southern skies and was the sight that Ian Shelton, a Canadian astronomer working at an observatory high in the Andes, first photographed and was thereby credited with the discovery. What is new this time is that we can see with other eyes: underground detectors captured some of the neutrinos produced from the collapse.

Although astrophysicists agree about the general scenario of the collapse, there is still much argument about the detailed mechanism for ejection of the outer regions of the star. These uncertainties limit the amount of information that we can deduce about neutrinos coming from the supernova. They poured through the detectors in a span of less than 10 seconds. At first many people argued that this implied a small but non-zero value for the neutrino mass: a spread of 10 seconds after 170,000 years' travelling limits their relative speed and hence mass *if* one assumes that they come out instantaneously. However, the theorists are now less certain about this. The 10-second spread might be due to the intrinsic duration of the neutrino emission, in which case they all got here at the speed of light, which is only possible if they are massless. At the other extreme one can imagine possible, if unlikely, scenarios where slow neutrinos come out first and the fast ones later; the fast ones then gradually catch up with the slow ones during the 170,000 year journey.

More papers have been written than neutrinos have arrived! The major differences among them have been the assumptions about the details of the neutrino production in the supernova. The most general conclusion is that the (electron type) neutrino has a mass less than or comparable to the maximum currently allowed by laboratory experiments (which involve radioactive decay of tritium and the indirect study of the neutrino that implicitly emerges in this process).

So a major question being debated is: what are the details of the stellar collapse that produces the hordes of neutrinos?

When a star like this collapses some 10^{57} positively charged protons merge with negatively charged electrons, neutralising their electrical charges to form neutrons, which are massive and remain to form the neutron star, and neutrinos, which fly off. This happens in less than 1 millisecond (1/1000 second), effectively instantaneously. This is dramatic and sudden to be sure, but these neutrinos carry off less than 10 per cent of the energy. The remainder come from processes occurring in the hot conditions, such as electrons meeting and annihilating with antimatter counterparts, 'positrons' (identical to electrons in all respects but for positive electrical effects in place of negative). This annihilation emits neutrinos and antineutrinos (the antimatter counterparts of neutrinos). More than half of the neutrino and antineutrino emission occurs in the first second and the rest comes out over the next tens of seconds as the newborn neutron star cools down into the familiar cold neutron star or 'pulsar'.

The detectors on Earth are sensitive to the arrival of the *anti*-neutrinos, but are almost blind to the neutrinos. Now the millisecond burst produces only neutrinos and these aren't seen on Earth; we only capture the antineutrinos emerging during the next *tens* of seconds. Hence, in part, the difficulty in learning much about the (anti)neutrino masses from the span in their arrival times

– it could well tell us more about the relative times of their production than about their flight.

There have also been some puzzles, so far not totally resolved, in that a 'small' ('only' 90 tonnes) detector under Mont Blanc responded differently to the burst than did larger, 1,000 tonne, detectors in Japan and the USA. The Mont Blanc detector was designed to detect antineutrinos from collapses in our galaxy, not from more remote sources like the Large Magellanic Cloud. The huge detectors in Japan and USA were designed for quite different physics, namely the search for signs of decaying protons, and their suitability for extragalactic neutrino astronomy has turned out to be a serendipitous bonus.

The discovery proves that our theories of stars are right – stars can and do collapse and change into neutron stars. Previously this was a conjecture with circumstantial evidence, but capturing the neutrinos from the blast is like a smoking gun – we have caught Nature at it.

There is much excitement at the prospects for the future of neutrino astronomy. The supernova occurred in another galaxy, at least six times more remote than anything in our own. So when a supernova occurs in our galaxy we should in future detect the neutrino blast with ease.

Supernovae are not that rare. Astronomers detect them in distant galaxies regularly. In our own galaxy, a supernova occurs on average every 20 years or so. The galaxy is a huge place and we are forced to look across it, rather than in from the outside where the view would be clearer. The chances are that the supernova is hidden from view by dust and other stars, or too distant for the naked eye to see. Only occasionally does one occur in full view. It is as if we are in the middle of a cosmic minefield, with explosions going off all around but so far, thankfully, not too near us.

Figure 9.1 shows the sites of those seen in our galaxy in the last millennium; we are long overdue for another visible one. However, the galaxy is rather transparent to neutrinos, so neutrino bursts from 'local' supernovae should reach us even if the light doesn't. And now we will be able to see those neutrinos – the observation of neutrinos from the Large Magellanic Cloud proves that we can do

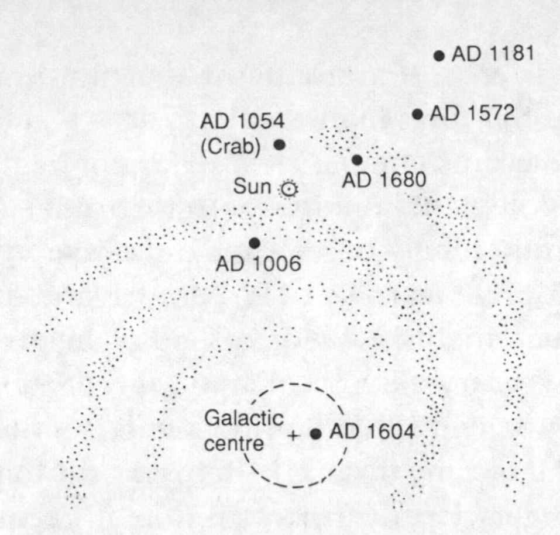

9.1 Supernovae in the Milky Way Location of supernovae in our galaxy relative to the Sun, spiral arms and galactic centre. There have probably been many more than these but obscured from view.

it. All we have to do is wait, and as the source of a signal in our galaxy is much nearer than the Large Magellanic Cloud so it should be much bigger. The message is that if a star collapses *anywhere* in our galaxy then, regardless of the luminous visibility of the supernova, we should observe it by its neutrinos.

Within a human lifespan we expect three or four such occurrences. With luck we will have the first by the end of the century and then we will learn in detail what happens when stars collapse. Meanwhile we watch the development of the new supernova with interest.

Moreover, now we can see if our theory of neutron stars is right. For as the shock wave rushes out into space it will thin and reveal the neutron star that we believe has been formed at the centre. Possibly, if enough mass is concentrated there, it will have collapsed further and become a black hole. We will know the answer in a few years; all we have to do is be patient.

Did a Supernova Kill the Dinosaurs?

According to current wisdom there do not appear to be any candidates for future supernova bursts in our immediate neighbourhood. Maybe this is because all suitable candidates have already erupted. There is some evidence that a supernova 5 billion years ago caused the formation of the solar system. Did a more recent blast kill the dinosaurs?

We know how elements are cooked in stars and so can calculate how much iridium should be ejected in an average supernova blast. In Chapter 5 we learned how Luiz Alvarez measured the amount of iridium in the Italian boundary layer between the Cretaceous and Tertiary geological strata. From this one can deduce how much was deposited worldwide. It is a vast amount, so much in fact that if it came from a distant supernova then that star must have been very near, at most 0.1 light years distant (the nearest known star of any sort currently is 4 light years away).

The chance that a supernova explosion occurred so near to us within even 100 million years is less than one in a billion. With such a low probability a nearby supernova is a one-off event. It either happened or it didn't. There are experimental tests that you can do and they prove that it didn't.

The trick is to use the fact that not all atoms of a given element are alike. Electrically neutral atoms contain electricity within them. Negatively charged electrons whirl around a positively charged compact nucleus containing positively charged protons. A single proton has as much positive charge as an electron has negative charge. So neutral atoms contain the same number of electrons as protons. It is this number that determines the chemical character, identifies which element it is. Thus hydrogen, the simplest, contains one electron and one proton. Helium, the next simplest, contains two electrons, while uranium contains 92.

A compact nucleus containing many positively charged protons would be highly unstable: 'like charges repel'. Nuclei survive because of a stabilising agent – electrically neutral particles called neutrons. Atomic nuclei contain protons and neutrons. Adding neutrons to a nucleus changes its mass but does not change the chemical identity. For example, U-235 and U-238 are two 'isotopes' of uranium; each contains 92 protons – the identifier of uranium – but they have different numbers of neutrons in their nuclei. U-235 contains 143 neutrons (making a total 235 of neutrons and protons), while U-238 contains 146 neutrons and so 238 neutrons and protons.

When elements are cooked in stars neutrons and protons are fusing together to form the nuclear seeds of atoms. A star will produce several different isotopes of any given element, the actual mix depending on the conditions within the star. Neutron stars, as their names suggest, provide copious neutrons and easily form heavy isotopes rich in neutrons. Highly evolved stars, on the other hand, tend to provide fewer neutrons and so the mix of isotopes is different. Astrophysicists can predict what abundance of elements and what mix of isotopes should result from various types of star, in particular a supernova.

A supernova explosion should send out iridium and also plutonium. There's plenty of iridium in the boundary clays, as Alvarez found, but no traces of plutonium at all. This was the first hint that a supernova was not the cause.

Iridium has two stable isotopes, iridium 191 and 193. Different supernovae will produce these two isotopes in different relative amounts. The solar system was formed from a primaeval gas cloud that was fuelled by the explosions of countless supernovae and the overall mix ended up with about twice as much of the 193 isotope as the 191.

Alvarez had discovered that the amount of iridium increased dramatically when he compared the boundary layer and the other rocks (page 66). Yet the ratio of the two isotopes is the same in both. A modern supernova would eject its own particular ratio of the iridium isotopes, which would be most unlikely to coincide with the particular value that the solar system contains. The two to one mix is the fingerprint of material from the solar system.

So although there are still arguments as to whether it was earthly or extraterrestrial, there are some things that we can be sure of. The disaster was not caused by a collision with some nearby star as we passed through the galaxy's spiral arms, nor was it due to a supernova.

10

A Universe of Galaxies

How Far Is It to Betelgeuse?

On a small lump of rock orbiting an insignificant star groupings of molecules have organised themselves in such a way that they have consciousness. They are aware of the awful gulf that separates them from the nearest star, the constellations and remote galaxies. Peering out from under a blanket of air they can measure how far away are those fellow partners in the ultimate joke. To be able to determine the make-up of those distant lights is one of the great accomplishments of human culture.

The ancient Greeks knew of the Milky Way but not what it consisted of. Not until the seventeenth century, when Galileo turned his telescope at it, did it reveal itself as millions of stars and misty patches or 'nebula'. Nebulae look similar to comets at first glance, so much so that many people thought they had seen Halley's comet in 1986 when in fact they had been gazing at nearby nebulae. Charles Messier in 1771 listed over 100 nebulae so that comet hunters wouldn't be confused by them. They are named after him, listed M (for Messier) in his catalogue – M1 (the Crab Nebula), M2, M3 and so on, like British motorways.

But what are nebulae?

Most people at that time thought that the nebulae were gas and dust in our neighbourhood. The philosopher Immanuel Kant, however, thought differently. He suggested that some of the nebulae are in the Milky Way but that others, like the beautiful spiral ones, are distant clusters of stars like our own. Nobody took much notice, not least because there was no way of testing his idea at the time. The first key to unravelling the universe came with the invention of the spectroscope. This split light into colours, like a prism but so fine that the Sun's spectrum, for example, was seen to be interlaced with hundreds of black lines. In the laboratory scientists discovered that each chemical element, when heated, gave out a tell-tale light, which a spectroscope analysed like a detective reading a fingerprint. Comparing the laboratory spectrum with that from the Sun revealed that the Sun contained hydrogen, iron and sodium along with other elements. Suddenly the dream of the ages became reality: we could tell what stars are made of.

William Huggins – a wealthy London chemist and keen astronomer – analysed the light of many stars in the mid-nineteenth century and then started on the nebulae. He found two types. Some were gas but others had spectra like the Sun, suggesting that they consisted of stars.

Now the problem was whether the stellar nebulae are inside or without our galaxy. Astronomers in the nineteenth century had little idea of how far away the stars are.

We know how far away the Sun is (p. 24); a ray of light makes it in just over 8 minutes. This mean distance to the Sun is called an 'astronomical unit' and is the baseline for measuring the distance to nearby stars. In six months' time the Earth will be on the opposite side of the Sun. It will be *two* astronomical units of distance away from its present position.

The view you have of the stars today and in another six months will be slightly different. As your left and right eyes in combination give you binocular vision and a sense of distance, so does the six-month interval give you a 'left'- and 'right'-eyed view of the stars. Nearby stars will be displaced by parallax relative to the more distant ones. Knowing how far the baseline is between January and July we can determine how far away a star is by its slight movement against the background (see Figure 10.1).

Our nearest stellar neighbours are Proxima Centauri and its more brilliant neighbour Alpha Centauri, visible in the southern hemisphere. The amount of parallax shows that Alpha Centauri is

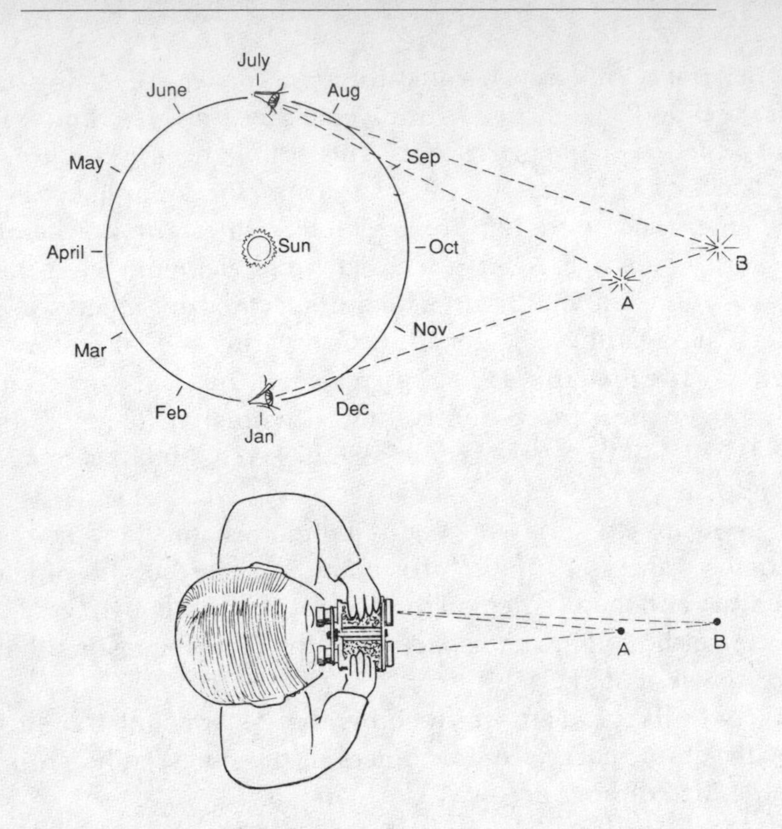

10.1 Binocular vision and parallax The human's right eye sees object A and B aligned whereas the left eye sees them separated. This parallax enables us to deduce that B is more remote.

The Earth orbits the Sun. In January, stars A and B are aligned ('right eye'); in July, we have the 'left eye' view and can tell that star B is more remote than star A.

275,000 times more remote than the Sun. A flight in Concorde would take 2 *million* years; light takes 4½ years. To express the distance in miles would be incomprehensible, so instead we say that it is 4½ light years distant.

We want to know not only where stars are 'now' but where they are moving, for things will not always be as we find them today. A star's motion can be broken down into two components – the 'radial motion' along the line of sight and the 'proper motion' at right angles to the line of sight.

The radial velocity is found by measuring the spectrum of light from the star, deducing what elements are giving the characteristic

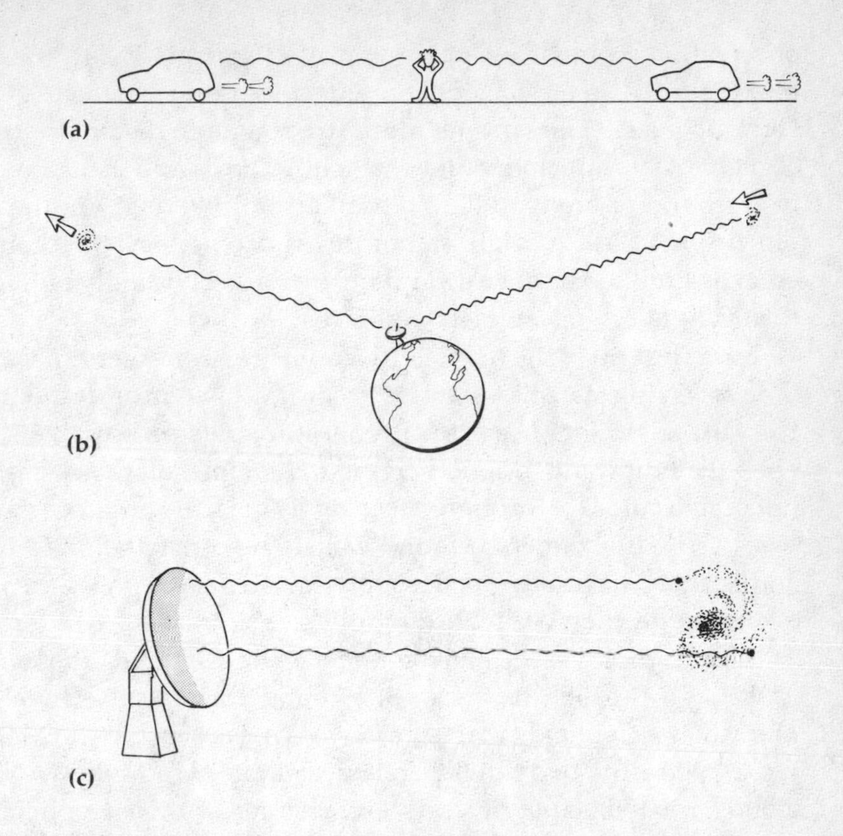

10.2 The Doppler shift (a) Car siren rushing towards or away from the listener appears to have higher or lower pitch respectively.
(b) Similarly a shift occurs in the colour of light. The galaxy on the right is approaching and its light waves are compressed – blue shifted. The one on the left is receding. Its light is stretched out to the red.
(c) A telescope can even detect the rotation of a galaxy from the different red or blue shifts of the spectrum of light coming from stars in different parts of the spiral.

colours and then comparing with the spectra that such elements give on Earth. The wavelength of the spectral light is Doppler shifted in a similar manner to the familiar shift in the pitch of sound as a car horn wails coming towards you, and dips as it departs: the spectrum of light is shifted to the red for a departing star and towards the blue as it approaches (see Figure 10.2). Radial velocities can be measured accurately for stars and entire galaxies out to the remotest reaches of the cosmos.

To measure the proper motion requires decades of careful

observation, comparing pictures of the star relative to the deep background. This is so subtle an effect that only the nearest stars show up a detectable proper motion. Stars more distant than 20 or 30 light years are too remote to show any motion against the background or any parallax as the Earth orbits the Sun; the third dimension is lost. A bright star or the Andromeda Nebula could be as far as 1 million light years and intrinsically blindingly radiant, or relatively faint and only 100 light years away.

So at the end of the nineteenth century astronomers spent their time collecting stars like stamps, recording spectra and colours. At Harvard, Henrietta Leavitt was collecting data on variable stars – ones that dim and brighten regularly. Some of these are very noticeable and go through their cycle in a few hours, days or weeks. The first one to be found was in the constellation Cepheus and so they have been collectively known as Cepheid variables even though they occur throughout the sky.

Leavitt discovered that the *brighter* the Cepheid, the *longer* it took to go through its cycle. She found this because by chance many of the Cepheids that she was looking at were in the Magellanic Clouds – the two satellite galaxies of our own Milky Way. The clouds are 150,000 light years away so all of their Cepheids are equally distant from us, give or take a few per cent. Confusing differences in brightness caused by varying distance (the plague of stars in our galaxy) were washed out. So Leavitt was able to show that the time that Cepheid took to go through its cycle tells its intrinsic brightness. Comparison with apparent brightness reveals its true distance.

Astronomers measured the Sun's drift among its fellow stars and managed to estimate the distance to nearby Cepheids. This gave an absolute distance scale and at last the third dimension had been added. At short distances parallax was the answer; further out the Cepheids provide the yardstick.

Now we come to the first decade of this century and meet Harlow Shapley at the Mount Wilson Observatory in California, housing what was then the world's largest telescope with a 60 inch diameter mirror. He was looking at Cepheid variables which are rather common in globular clusters – groups of as many as 100,000 stars of amazing beauty. Shapley collated data on the Cepheids in them and then began to map out a three-dimensional impression of the globular clusters in the galaxy.

Over 100 globular clusters are visible from the Earth. In a dozen of the closer ones he identified Cepheids and measured distances. In some of these nearby globulars he isolated bright red giants and supergiants and systematically compared their apparent brightness with that of the Cepheids. Soon he had gathered enough data to gain a feel for the intrinsic brightness or 'absolute magnitude' of these giant stars.

In distant globular clusters the bright giants are about all you can see; the Cepheids are too faint to study. But with these giants as 'standard candles', reference scales of brightness, Shapley began to chart the distances of globular clusters deeper into the galaxy.

By 1920 he had a three-dimensional map of the globular clusters. What it showed astonished him.

The globular clusters aren't spread out randomly around the Milky Way. They are concentrated in a vast sphere as if they are components of a 'super globular'. The centre of this sphere is 30,000 light years from us in the direction of the constellation Sagittarius. Shapley made an inspired guess: the centre of the globular sphere is the centre of our stellar system, the Milky Way. Ptolemy had put the Earth at the centre of the heavenly sphere; in 1543 Nicolaus Copernicus had overthrown this and proposed that the Earth orbits around a central Sun. Now Shapley suggested that the Sun isn't at the centre of the galaxy but is instead far out in the remote suburbs. Once it has been suggested it becomes almost obvious. Compare the Milky Way viewed from northern and southern skies. It is relatively faint when we look out in North America and Europe, but in South America and Australia you see rich brightness as you look through 90 per cent of the disc and into the heart towards Sagittarius.

One thing that Shapley overlooked was the dimming effect of interstellar dust. He thought that dimness was due to remoteness and estimated that the galaxy is 200,000 light years in extent. In reality it is smaller, nearer 100,000 light years; the dimness is due to the fog. It was probably because of this overestimate that he failed to discover that ours is but one in a family of galaxies. The Magellanic Clouds are satellite galaxies of the Milky Way 150,000 light years away – 50 per cent more than the extent of the parent galaxy. Shapley's overestimate of the Milky Way's size had the Magellanic Clouds inside rather than separate entities.

The discovery of a Universe of galaxies was made by Edwin Hubble working at the new 100 inch diameter telescope at Mount Wilson in the 1920s. With this he was able to resolve spiral nebulae, like Andromeda, into stars. The proof that they were galaxies came when he was able to photograph 50 Cepheid variable stars in a galaxy visible in southern skies and deduced that it was several hundred thousand light years away. This was so far away, several times more remote than a voyage across our entire galaxy, that Hubble was relieved to discover that deep in the cosmos the same physical laws were at work. This enabled him to carry on outward.

The two largest spiral nebulae visible from the Earth are Andromeda and M33. M33 is oriented face on to us and Hubble photographed it with the 100 inch telescope over two years. He resolved it into stars and identified 35 Cepheid variables. In February 1926 he wrote his report: M33 is 2.4 million light years away. This was far out compared to anything previously known. Without doubt it was a spiral galaxy separate from our own. And at Christmas in 1928 he published a paper on what he had discovered about Andromeda. The million light year barrier had been crossed. So for the first time, and within the memory of senior citizens still alive today, human culture had identified its place within the cosmos; the Universe is divided into galaxies separated by vast spaces void of stars.

It had taken a long time to get from the Earth as centre of the Universe, to us orbiting the Sun in a remote corner of the galaxy. And now it is no longer *the* galaxy but *a* galaxy; one of a billion stars in one of a billion galaxies.

Once Hubble had realised that the Universe is built of island galaxies in the vast sea of space he quickly discovered that the Universe is expanding, evolving in time, that the galaxies are rushing away from one another.

This is an insight of truly cosmic significance for it teaches us about the origins of the Universe. If the Universe is expanding then imagine stopping the clock and playing the film in reverse. In the past everything must have been much closer than it is now. About 10–20 billion years ago all of the material in the Universe must have been crammed into a volume smaller than a clenched fist. The explosion out from this dense ball is called the 'Big Bang'. Genesis occurred in an explosion 20 billion years ago and has been expanding ever since. None of this was known; then Hubble knew it and now it is part of human wisdom.

He came upon this by accident. Initially he had wanted to know how fast the Sun is moving around the galaxy. If you are on a roundabout you can tell how fast you're moving by looking fixedly at distant reference points. Use distant galaxies as the fixed references and see if the Sun is moving towards, away or across them, then work out the carousel.

A few nearby galaxies showed the random motion expected. More dramatic was a marked tendency for distant galaxies to be rushing away, and faster the more distant they are from us. Hubble worked out their speed by breaking down their light with a spectroscope and seeing how much it was shifted compared to the catalogue of colour-fingerprints that was by then well known.

Hubble charted clusters of galaxies deep into space and by 1934 improved photographic materials enabled him to take pictures of galaxies that were like dots, more numerous than the foreground stars. Hubble and his team found distant galaxies that are rushing away at one seventh the speed of light. The current record is held by galaxies so remote that their light has been travelling across space for 10 billion years until some of it chanced upon a telescope on an insignificant planet. When we look at such images we are looking back towards the start of the Universe.

The question that this raises is whether the Universe will continue to expand or whether it will eventually collapse under its own weight. That is a question currently taxing the cosmologists and also particle physicists who recreate the violent conditions of the early universe in earthbound laboratories.

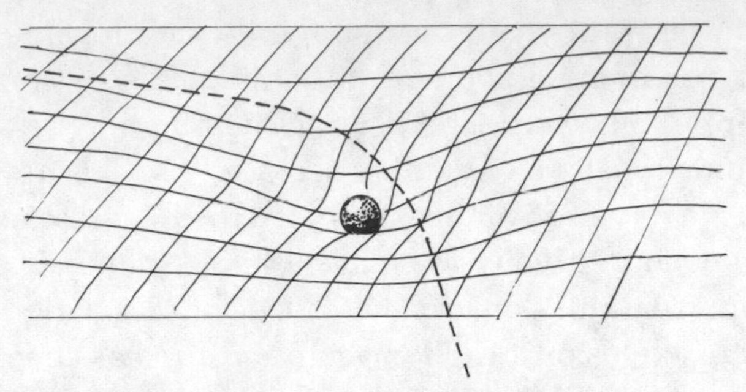

10.3 Gravity warping space–time A massive body warps space and time. The dashed line shows the path that a ball rolling across this imaginary space–time surface would follow. The deflection around the warp is interpreted as resulting from the force of gravity.

The End of the Universe as We Know It

Albert Einstein made a blunder when he wrote down his general theory of relativity – the theory of gravity that superseded Isaac Newton's great seventeenth-century work. Einstein got everything right but for one thing.

In Einstein's theory of gravity, space and time are subtly intertwined. He viewed space as warped by the presence of things; when we meet a twist it tips us off our straight path. The apparent shove is what we call the force of gravity. Step off a high bridge and it is the spacewarp caused by the Earth that pulls you down.

An old analogy may help. We live in three dimensions but suppose that it were two – that we were 'flatlanders'. Represent this by a taut two-dimensional rubber sheet. If you drop a pea on it, it will make a small depression; a heavy ball will make a deep warp in the elastic sheet. Roll a ball across the sheet and it will curve around the warp. A flatlander would say that a force – 'gravity' – attracted it. We, with our greater vision, say that it is the warped space that was responsible – the warping caused by the object's mass.

In the real universe, space appears three-dimensional and masses – the Earth, Sun and galaxies – warp space in a fourth dimension. This may be hard to visualise; in fact it is even harder! Einstein says space AND time are warped. The Sun bent space around it and we are circling in this warp.

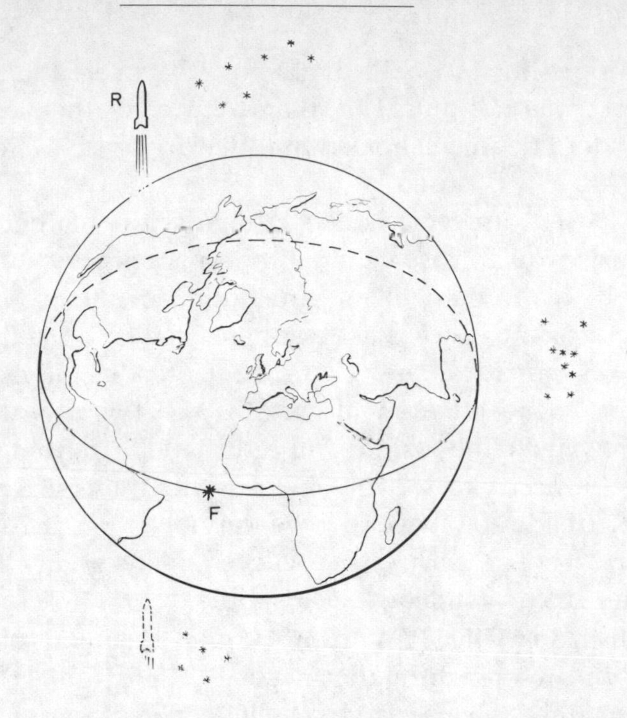

10.4 Round trips in higher dimensions Flatlander F can experience only two dimensions, circles the globe and finds itself back at the starting point!

The rocket ship R contains intelligence – we know of 3D. We set off north towards the northern stars, eventually to return via the Southern Cross having circumnavigated the universe – curved in the fourth dimension.

Massive objects warp space more – galaxies warp it more than you or I do. But how is space-time warped on the scale of the whole Universe? The long-term future of the Universe depends on the answer to this.

This is where Einstein made his blunder.

At the beginning of the century astronomical data were much more sparse than today and of poor quality. They appeared to suggest that the Universe is static, permanent and unchanging. Einstein's theory, however, implied that the Universe is evolving. This conflict worried him and so he introduced an extra piece (known as the 'cosmological constant') into his equation to fit with what Nature seemed to require. Present data show that it is not needed. The Universe *is* expanding and if Einstein hadn't tinkered with his pristine theory he might have discovered this for himself. In fact it was not until 1922 that a Soviet mathematician named Alexander Friedman showed that the Universe is evolving and is either 'open' or 'closed'.

A 'closed' universe is like a huge self-contained black hole. There is so much mass within it that space is curved right back on itself. If this is hard to imagine, think of the two-dimensional analogue again: space is like a sphere, in which the flatlander will crawl off eastwards and circumnavigate the globe. In the real universe, head off from the North pole towards the pole star, circumnavigate the cosmos and return to Earth from the direction of the Southern Cross. In other words the Universe is finite; if there is an 'outside' or 'beyond', we cannot reach it. In the 'open' universe, by contrast, space curves away gently forever. The two-dimensional analogue is a saddle.

Which do we live in? We don't know. It depends on the average density of matter throughout space. If it is more than a critical amount, the universe is closed; less, and it is open. Einstein's theory allows no other possibility. Some theories of the Big Bang (the so-called 'inflationary universe') require that the density is exactly equal to the critical amount. If we could detect all the forms of matter in the heavens we could predict whether the universe will collapse or expand for ever. But there is a lot of stuff too dark to see and the big question today concerns the nature and abundance of this 'dark matter'. The modern 'superstring theory' (Chapter 12) suggests that there could be an entire dark universe operating in parallel with our own! We need to know if this dark stuff is there or not. How can we hope to see 'invisible' stars?

What Is Hiding in the Dark?

Go around a corner too fast and you will leave the road. Rush around the Sun or the galaxy too fast and you will spin off into deep space. Planets feel less gravitational pull the further away they are from the Sun, and their speed in perpetual orbit is correspondingly less. Each of us has settled at a speed such that we fall in and centrifuge out in balance.

When we look at spiral galaxies, however, the stars appear to rotate too fast to survive; the visible stars do not have enough mass to hold the systems together by gravity – yet survive they do.

Clusters of galaxies move around one another. Here again they move too fast and yet manage to continue in their cosmic dance. Something else must be holding them together – a ubiquitous ectoplasm too dark to see, permeating the galaxies, utterly invisible and with enormous mass. No one yet knows what this stuff is but astronomers are certain that it is there. The galactic dance shows what tune Nature is playing and there is more in the orchestra than we have detected so far.

Roughly 90 per cent of the mass of the Universe must be made up of this dark matter. The ordinary visible matter, shining in the heavens, picked up in X-rays, infra-red and radio telescopes, is a flotsam in a sea of invisible stuff. The dark matter is controlling the show.

There can't be much of it in our immediate neighbourhood because the planets orbit the Sun on schedule. However, it hasn't always been like this. The most famous successful prediction of dark matter was that of the planet Neptune. 150 years ago Uranus was the outermost known planet and its orbit didn't 'obey the rules'. In 1843 John Adams in England predicted that another unseen planet existed and told the astronomers where to look for it, but he was a young unknown and no one took much notice. Three years later Urbain Le Verrier in France independently had the same idea and a squabble ensued following Neptune's discovery that year as to who should get the credit.

The discovery of Neptune was a great success: the prediction of dark matter from its effects on the visible – like H. G. Wells' invisible man giving his presence away by jostling the crowd. Le Verrier tried to repeat his success in the case of the innermost planet Mercury. This too didn't obey the rules and Le Verrier suggested that there is another planet, which has been named Vulcan, near to the Sun. None has been found, and today we know that the cause of Mercury's anomaly is the failure of Newton's theory of gravity in strong gravitational fields like those near the Sun. In such circumstances Einstein's theory of general relativity applies instead and describes the orbit of Mercury perfectly.

Some physicists have tried to avoid dark matter altogether by suggesting that Newton's laws break down on the galactic scale, hence the 'anomalous' behaviour of the outer regions of galaxies. However, other physicists claim that this idea runs into problems with an empirical relation between the luminosity and velocity of galaxies. Current opinion tends not to favour this solution to the conundrum.

There may be a little dark matter about in the outer reaches of the solar system that is responsible for anomalies in Neptune's orbit. (As I remarked on page 48, tiny Pluto doesn't really fit the bill and Planet X, as yet unseen, is the favourite.) Incidentally, this is an amazing testimony to the precision of the measurement of Neptune's motion. Neptune orbits so slowly that since its discovery 150 years ago it has not yet completed an orbit!

The stars in distant galaxies behave as if the dark matter forms haloes on the periphery of the galaxies. Being immersed in our own dust-filled galaxy it is sometimes more difficult to learn about things in our own Milky Way than in remote galaxies where we can peer in from the outside, watching the whole scene. But we can get indirect hints by watching the dwarf galaxies, near neighbours of ours, and seeing how their stars behave – they are near enough to us to wallow in any dark halo that we may have.

These dwarf galaxies have as few as a few hundred thousand stars. They lie in a plane along with the Magellanic Clouds and our huge Milky Way, suggesting that we are all connected, the remnants of an original protogalaxy in the early Universe. These romantically named clumps – Sculptor, Draco, Carina and others – are so small that you would hardly notice them if you were looking in from Andromeda. So, in turn, if Andromeda has dwarf galaxies as small as these, we are unlikely ever to know much about them short of making a trip. This suggests that dwarf galaxies throughout the Universe could add up to a considerable amount of 'dark' matter.

Unfortunately, it is very difficult to see our own dwarf galaxies at all, let alone make precise measurements on their constituent stars. Some observers claim one thing, others another.

However, the great Milky Way and our giant neighbour Andromeda do seem to be misbehaving. We're rushing towards one another, closing at about 100 miles every second. Now we don't need to worry about this, as it will take several billion years for us to collide even if we are on collision course, but the fact we're doing this at all causes some concern.

It is possible that we are just like ships passing in the night; two great galaxies that are on independent journeys through the cosmos. But it seems rather a fluke if it is so. Given the immensity of space that either of us could have been in, these two galaxies seem to be connected, bound in mutual orbits around one another as are the planets around the Sun. If we have been inextricably linked since our formation in the post Big Bang, then we shouldn't be closing so fast. Lots of dark matter must be around to provide the extra tug; the dwarf galaxies alone are trifling.

The consensus is that dark matter abounds, but we are not sure about the details of how much of it there is and where. It is important to answer this question because the Universe is an expanding, ever-changing, living thing and it will be the dark matter that determines its eventual fate. If there is a lot of matter around, its gravitational pull will weigh the Universe down, slowing its outward flight to the point where it stops and collapses in a big crunch. If there is less than this critical amount then the Universe will expand forever, the stars and entire galaxies will exhaust their fuel, matter will erode away leaving only electrons and radiation. We cannot say with certainty which way things will turn out, because the Universe appears to be very close to the critical dividing line between collapse and continual expansion.

No one yet knows what this dark matter consists of. Astrophysicists have had to turn to particle physics for ideas. Neutrinos could be the answer. If the Universe began with a hot Big Bang, then theory predicts that there should be about 100 neutrinos in every cubic centimetre of space – 100 million times the density of protons, which provide most of the visible matter in the stars. The visible galaxies would be merely islands in a vast sea of neutrinos.

If neutrinos have a little mass they would have been moving at almost the speed of light during the first 10,000 years after the Big

Bang, rushing outwards with the expanding Universe. As the Universe cooled, the massive neutrinos would have started forming clusters under the influence of their mutual gravitational attraction. These conglomerations would have covered the entire Universe. Local instabilities within them would have formed the core of galactic clusters out of which individual galaxies condensed. In the jargon of astrophysics this scenario is called 'top down' because the small galaxies emerge from the large clusters. However, the galaxies in the real universe are more patchy than computer simulations of the top down mechanism would lead us to expect.

Modern theories of matter and the natural forces are remarkably successful in describing experiments performed with subatomic particles at particle accelerators. Indeed, many theorists have been encouraged to apply these ideas to phenomena at present out of reach of laboratory experiments but accessible elsewhere in the Universe. One example is that they make profound statements about the nature of the early universe in the hot Big Bang and the origins of matter in that epoch.

However, not everything is perfect. There are some technical mathematical problems that many theorists believe can be solved if massive particles, as yet undiscovered, exist. These have exotic names like axions, photinos, Higgs bosons and monopoles, and experiments are currently looking or being planned to search for them in the near future. Until we find proof that they exist these particles are just postulates – but if they are real then they will have been formed in the Big Bang along with the particles that eventually seeded our atoms. Although uncommon today they may have left fossil imprints in the way that the Universe developed.

Some of these particles are predicted to be much heavier than protons, thousands maybe billions of times so. Some may even be stable. In the first moments of the Big Bang when the Universe was unimaginably hot, much more so than in any star today, any massive particles would have flitted around like everything else. But the Universe cooled faster than a cup of coffee in Siberia; the massive particles froze still within seconds. As the other lighter particles clustered to build the galaxies, the stars and eventually the matter that we are familiar with today, their sluggish cousins would have gravitated together, forming conglomerates that may be in stars like our Sun or may be on their own. Indeed, one theory is that these bulky lumps were the seeds that ensnared the fast-moving lightweight electrons, neutrons and protons forming clusters on all scales, from clusters of stars to individual galaxies or groups of galaxies. The small ones would have developed first and then coalesced later into larger clusters. This is known as 'bottom up' and seems to be what the actual universe is like.

So the motion of stars within the galaxies suggests that there is unseen dark matter around. The distribution of the galaxies throughout the Universe seems to imply that this dark matter consists of massive particles not yet seen in experiments at the world's high-energy accelerator laboratories. The future not just of the Sun, but of the galaxies and the entire Universe depends on unknown forms of matter. The future of astrophysics and cosmology is coming to depend more and more on particle physics – the study of the basic constituents of matter.

Recent discoveries about the behaviour of these fundamental particles turn out to have surprising implications for the future of large-scale structures, including you and me.

Part IV

The Heart of the Matter

11

How Stable is Matter?

All the rivers, lakes and oceans froze with a 'great *vvarroomph*' as Kurt Vonnegut's mad scientist dropped some molecules of 'ice-nine' into a stream. Ice-nine was a (hypothetical) form of water, more stable than the familiar form, which froze at room temperature. In the story, ordinary water is metastable and changes into the stable form – ice-nine – when it encounters the minutest traces of it.

Put ice-nine in a whisky and soda and you would have instant scotch on the rocks; but don't drink it or the water in your body will immediately polymerise. If water were metastable then we would have a hazardous existence.

Thankfully that's science fiction. However, some theoretical physicists believe that the Universe might contain stuff similar to ice-nine, called 'strange matter', more stable than the stuff that we're made from. Atomic nuclei, the seeds of matter, are electrically charged and could attract strange matter like the north and south poles of two magnets grip one another. The stable strange matter would then devour conventional matter. Not just water but all atoms of everything would be vulnerable. Your world could literally collapse around you.

We are fairly confident about the Universe that we can see: we can reproduce details of it in the laboratory, watch it at work under controlled conditions, predict eclipses, build machines that can fly across space – all based on our understanding of the physical laws. We can look at the light coming from distant constellations and remote galaxies; the spectral fingerprints show that the same laws operate universally. We can even predict the death of stars, and when one does implode, as in the 1987 supernova, we can watch it and we find that it behaved as our theories implied it should. The message is that the visible Universe out there is made of the same varieties of particles as you and me, the ubiquitous electrons, protons and neutrons, but combined in different ways at different temperatures and pressures. This is a remarkable generalisation and must rank among the great achievements of the human intellect.

Having the basic ingredients available to us here on Earth, we can study them in the laboratory under conditions that they would encounter in the stars and see how they behave. We can even push them to temperatures prevalent in the first moments of the Universe's existence and learn about the origins of bulk matter. And when we do this we begin to see hints that the Universe that we know may be only a small part of the whole.

So far in this story we have been concentrating on the Universe in the large – the behaviour of bulk matter ranging from little stones in the meteor showers, through comets and asteroids, the life and death of stars and ultimately the collective motion of entire galaxies. As we have looked into these more closely we began to see hints that there are some gaps in our knowledge. What is going on deep inside the Sun? What is the dark matter that gives itself away by jostling the galaxies? Are these problems that are intellectually interesting but of no practical concern or might they pose hazards for us?

We met some suggestions as to what is going on. The particles called neutrinos coming from the Sun brought the message that something was untoward and the debate centres on whether it is the neutrinos that are themselves to blame or whether there are unknown massive particles, WIMPS, at the centre of the Sun. Maybe there are WIMPS throughout the cosmos and they are the dark matter. Whether this is the case or not remains to be seen but I'm recapitulating it here to bring out the change in direction that we are about to take in this tale. We are beginning to concentrate less on the bulk matter, more on the little particles that build it up. By watching neutrinos in the laboratory we hope to see if they behave anomalously or not, and hence whether they or something else is responsible for the solar neutrino problem. By smashing electrons or nuclear particles into each other, we can reproduce such concentrations of energy that it can congeal into new forms of matter, such as WIMPS if they exist. The basic particles of the Universe emerged from the intense heat of the Big Bang, and by recreating that heat in the laboratory we can see what Nature has on the menu.

Already there are tantalising and disturbing questions being raised as a result of this research. Are the basic particles stable or do protons decay, albeit very slowly, and thereby make the Universe erode inexorably? Space and time appear like a skeleton on which the living Universe evolves. Could space–time collapse? Could time stop and run backwards, jump about discontinuously, one region of space disconnect from another so that we became cut off, stranded on some huge cosmic analogue of a melting ice floe? We are all descendants from an initial Big Bang that erupted out of nothing, creating space and time with it; could this happen again within our present Universe? Why are there three spatial dimensions, or are there higher dimensions than those we are familiar with? Could they bubble up in your living room so that the familiar dimensions of up, forwards and sideways fragmented into some unimaginable foamy structure? Could there be other universes, unseen, operating in parallel with our own? Are there dark stars with their own planets in our immediate neighbourhood?

Many of these questions are currently being debated. A few years ago a list of questions like these would have been dismissed as no more than ideas for a science fiction novel. Today it is hard for the layman (and even many scientists) to tell which questions are clearly fictional and which are serious science. A paradoxical change is taking place at the moment. On the one hand we understand the Universe more deeply than ever in history, and have testable theories of how it emerged through to how it will die. Yet we are also becoming aware that the more we understand so the more bizarre are the possibilities for our ignorance. The Universe may indeed be stranger than we can ever know.

So we will now take a voyage into matter to see what it teaches about the fate of the Universe. Then we will look at the latest theories and learn of things that are stranger than science fiction.

The Core of the Cosmic Onion

The study of the Universe at large has been the province of astrophysicists, astronomers and cosmologists. They have concerned themselves with structures whose dimensions exceed light years in extent. At the other extreme we find the microscopic wonderland of crystals, molecules, atoms and subatomic particles. These so-called 'elementary particles' are the common building blocks of all known matter in the Universe. So to understand our origins and perhaps see clues to the ultimate fate of the Universe we need to study its smallest pieces.

Take a deep breath! You have just inhaled oxygen atoms that have already been breathed by every person who ever lived. At some time or other your body has contained atoms that were once part of Moses or Isaac Newton. The oxygen mixes with carbon atoms in your lungs and you exhale carbon dioxide molecules. Chemistry is at work. Plants will rearrange these atoms, converting carbon dioxide back to oxygen, and at some future date our descendants will breathe some in.

If atoms could speak, what a tale they would tell. Some of the carbon atoms in the ink on this page may have once been part of a dinosaur. Their atomic nuclei will have arrived in cosmic rays, having been fused from hydrogen and helium in distant, extinct, stars. But whatever their various histories, one thing is certain. Most of their basic constituents have existed since the primordial Big Bang at the start of time.

Atoms are the complex end-products of creation. Their basic constituents were created within the first seconds of the Big Bang. Several thousand years elapsed before these particles combined to make atoms. The cool conditions where atoms exist today are far removed from the intense heat of the Big Bang. So to learn about origins we have to see within the atoms, study the seeds of matter.

If you want to see what things are made from, you must look at them closely. You see things by shining radiation on them, such as light from the sun or a lamp bouncing off this page into your eyes. However closely you look you will not be able to see the carbon atoms in the ink of these letters. Magnify them as much as you like – it simply isn't possible.

The power of a microscope is not its ability to enlarge things; rather it is the ability to separate things that are very close together – its resolving power. To see atoms you must be able to separate one from the next. Visible light can't resolve distances smaller than about one-thousandth of a millimetre. There is a law of Nature which asserts that the smaller a thing is, the more energetic radiation you need to resolve it. I can't explain why it is this way – that is how Nature is. Visible light does not have enough energy to do the job.

This is where electron microscopes come in useful. By accelerating beams of electrons in a high voltage you can create a powerful enough radiation to resolve structures as small as atoms.

This is the world of high-energy particle accelerators, which reproduce the intense heat of stars in the laboratory and create feeble imitations of the Big Bang in small volumes of a few atomic dimensions. Huge machines, miles in length, accelerate pieces of atoms until these particles are moving near the speed of light. They then smash into atomic nuclei of material waiting at the end of the acclerator. These experiments show us the inner structure of the atomic nucleus in fine detail. We have identified the processes that fuel the stars as a result.

In 1967 at Stanford in California a 2 mile long accelerator of electrons began operation. Over 20 *billion* volts speed the electrons along an evacuated pipe. Starting their journey near the San Andreas fault, they dive under a freeway before emerging 2 miles distant into a huge hangar where concrete protects humans from the intense radiation. In the electrons' path sits a target of material whose inner details are about to be revealed.

The deep structure of matter is layered like an onion. Atoms consist of negatively charged electrons encircling a positively charged compact nucleus. The nucleus consists of positively charged protons and neutral neutrons.

The electron beam at Stanford is so powerful that it can reveal the fine details not just of atoms or of the atomic nucleus but of the

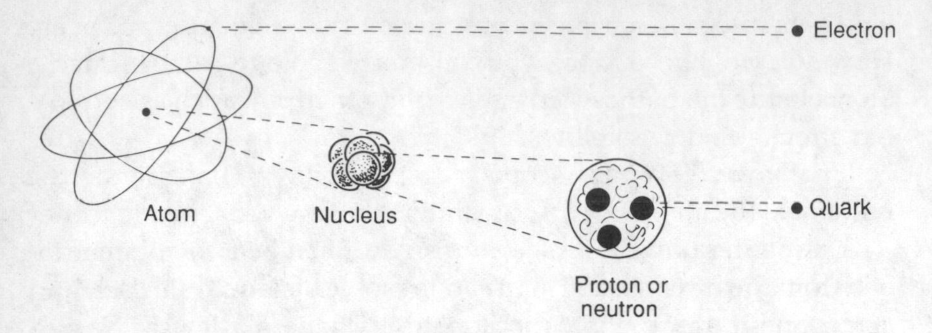

11.1 Structure of matter smaller than atoms Atoms consist of negatively charged electrons surrounding a positively charged compact nucleus. The nucleus is built from positively charged protons and electrically neutral neutrons which help to stabilise it. The protons and neutrons in turn are clusters of quarks. Present experiments have not found anything smaller than electrons and quarks and these are the currently accepted elementary particles of matter. The motion of electrons in the atomic periphery or the vibration of protons and neutrons in the nucleus can liberate energy in the form of light (photons) or electrically neutral neutrinos.

protons and neutrons that build up the nucleus. Here for the first time we could peer inside protons! And we found that these little particles have a detailed inner structure of their own. They are made of smaller pieces called 'quarks' (see Figure 11.1). (You can read the story of this voyage into matter and see pictures of the particles in *The Particle Explosion* listed in the Suggestions for Further Reading.)

To the best experimental accuracy yet available these quarks, along with electrons, appear to be the basic building stones of matter and were fashioned within one-billionth of a second of the Big Bang. They are the fossil relics of creation.

The discovery of the quark layer of reality may one day be perceived to be the crucial key in unlocking an entire new concept of the natural laws. It has changed the outlook in high-energy particle physics and made us realise that it teaches us not simply about the stuff of present matter, but also about its origins, *our* origins, in the Big Bang. It is making us aware of the possibility of new forms of matter that could pervade and disturb the Universe and suggests that the Universe could be a much more fragile thing than we had thought.

The Universe erupted from nowhere in the so-called Big Bang some 10–20 billion years ago. The matter that now fills the heavens, out beyond the limits where even the most powerful telescopes have yet peered, was in those first moments contained in so small a ball that you could imagine it all compressed within the dot at the end of this sentence.

Under these conditions everything would have been unimaginably hot. Heat has many forms, one of which is that the bits and pieces are all crashing into each other very violently. This is the sort of thing that the particle physicists are doing in the laboratory. By accelerating the basic particles of matter to high speeds, and then smashing them together, they are re-creating in the laboratory, in a small region of space, the sort of conditions that were present in the Universe in the moments following the Big Bang.

Ironically they are finding that it is relatively simple to describe the laws operating under these initial and extreme conditions. In the cool environment of the Earth a wide variety of phenomena occur and a whole variety of laws are required to describe them. But in the original hot Big Bang it seems that a uniformity existed, described by a single law. It is this discovery of a possible unified theory that is exciting the physicists at present and allowing them to work out models of the entire life of the Universe.

Every culture has had its favourite theory of the Universe; some of these are the foundations of whole religions. Bishop Usher even calculated the date and time in 4004 BC when, so he claimed, Genesis occurred. The important new feature in our modern theory is that it can be tested in the laboratory. If it fails, it is useless and must be discarded. So far it has stood up to the test. It may even be correct. This is what it portrays for our origins – and our future.

The Origin of Matter

The heat of the original Big Bang still survives, bathing the entire Universe in the dying embers of its glow. The Universe became transparent to this radiation some 700,000 years after the Big Bang. Since that time matter has been clumping together into stars and galaxies. Meanwhile the radiation has continued to expand and cool. It is now −270 degrees Centigrade, 3 degrees above absolute zero. This '3 degree microwave radiation' is the ambient temperature of the background Universe.

This radiation wafts on the particles of matter forming the large-scale structures in the heavens. These are built ultimately from electrons and quarks (the seeds of atomic nuclei), and little else, all of which fused in the initial heat of the Big Bang when the temperature was billions of degrees. Energy congealed into particles and antiparticles – entities with the same mass but opposite electrical charge to the corresponding particle. Thus the negatively charged electron was created along with its positively charged counterpart, the positron. Similarly quarks, which later clustered together forming protons and neutrons, were formed along with antiquarks (which make antiprotons and antineutrons). Electrons encircling protons and neutrons make the atoms of matter; positrons encircling antiprotons and antineutrons make up atoms of antimatter.

Here we meet one of the mysteries of Nature.

Gravity attracts everything inexorably to everything else. It holds our feet on the ground. That our bodies stay together, rather than falling in a pile of dust to the floor, is in part owing to the intense electrical forces within our atoms. The existence of bulk matter is the result of attractions between oppositely charged particles, in particular negatively charged electrons and positive protons in the neutral atoms.

It is an important fact of life that atoms containing the same number of electrons and protons are overall electrically neutral. The negative charge on the electron exactly balances the positive on the proton. But suppose Nature had switched the charges around such that the electrons were positive and the protons negatively charged. Atoms would still be neutral and bulk matter would appear exactly the same as it does in the real world. If we were made of this 'antimatter', physics would be identical to that in reality, where we are made of matter.

The heat of the Big Bang must have made particles and antiparticles in equal quantities – we can even reproduce the processes in the laboratory and photograph them. Matter and antimatter mutually annihilate when they meet, yet somehow an excess of matter survived this cataclysm. This now comprises the visible Universe.

If antimatter existed in bulk there would be an interface between regions of the Universe made of matter and regions made of antimatter. At this interface there will be violent annihilations going on continuously and intense bursts of gamma radiation emitted as a result. Astronomers could not fail to see these, yet none has been found. It seems that we are safe from one form of (literal) annihilation at least. Lumps of antimatter are unlikely to come raining down from the heavens, frying the Earth into gamma rays.

Here on Earth it is a slow process making large numbers of antiparticles. In several years at CERN, the European Centre for Nuclear Research in Geneva and the world leader in making antiparticles, they have made less than a billionth of a gram. So it would take several thousand years to make enough for a single antimatter bomb, even supposing you could preserve the antimatter that long in the hostile environment of matter all around it. So we can safely discount media stories of antimatter bombs – the so-called 'ultimate weapons'.

Where did all the antimatter go to?

Modern theories suggest an answer to this. While studying the behaviour of subatomic particles the physicists noticed that some particles known as 'kaons' appear to distinguish between matter and antimatter; the kaons of matter and those of antimatter behave in slightly different ways. Our understanding of this suggests that during the Big Bang the protons of matter were put together slightly faster than their antimatter images – the antiprotons. This has led to a net excess today – the galaxies of matter are the remnant of a *small* imbalance between particles and antiparticles. The Big Bang must have been very very big!

The theory of this implies that if protons were 'put together' in this way then they can also fall apart. Protons will be unstable.

Now, as the nuclei of every atom in our bodies contain protons, you might wonder how it is that we are still here. The answer is that the protons are *nearly* stable, their half life is predicted to be around 10^{32} years. This means that if you have a large collection of them, half will have decayed after that amount of time has elapsed. As the Universe has only been here for 10^{10} years – a mere ten-thousandth part of a billionth of a billionth of the proton's half life – most protons are still alive and well.

Even with this astonishing stability, physicists still hope to be able to test this theory and catch sight of the rare proton that dies before its time. In a large swimming pool filled with water there are so many protons – in excess of 10^{32} – that statistically one or two may decay in a year. Physicists have built huge tanks of water and surrounded them with detectors in the hope of catching the flash of light as a proton dies. The tanks are deep underground, away from stray influences such as the perpetual rain of atomic particles hitting the Earth's upper atmosphere which could mimic the flash of a decaying proton. There are tantalising hints but no clear evidence yet.

If they detect a decaying proton with certainty they will have gained the first sign that matter is eroding and hence seen the first glimpse of a dying Universe. In the century of your life, one or two protons may decay in your body, but you won't notice. In the deep future this erosion of the seeds of matter will become more noticeable. It is a cancer eating away at the fabric of the Universe.

If the Universe expands for long enough, then matter will begin to die off as the protons decay. However, if there is enough mass around then the Universe will slow and begin to collapse under its own weight. Either of these possibilities has consequences for our own survival prospects. The collapse into a hot rerun of the Big Bang, but in reverse, is a daunting prospect. The cool expansion into eternity with the constituents of matter eroding away looks bleak, but may be survivable!

Road to Eternity

In an open universe, or even a closed one if it lives long enough before the collapse begins, the ultimate demise of matter is a certainty. Even if protons are stable on a time-scale of 10^{32} years, they will eventually decay courtesy of the intervention of quantum mechanics. In a nutshell, quantum theory implies that if you wait long enough, anything will happen. In the case of the proton it implies that once in a blue moon a proton will spontaneously form a black hole and disappear. Blue moons occur on average every 10^{54} years – on the scale of eternity this is a small time-span.

As matter is certain to decay, we will have to change form if we are to survive. The matter that constitutes life as we know it will all die out and there will be nothing left to procreate succeeding generations. One possibility is that we discover a way of stopping or reversing the process. As this will require as yet unknown laws to be discovered, we have no idea if this is possible in principle and it is not realistic to speculate on it here.

These insights have come as a result of experiments concerned with the ultimate heat that was prevalent in the Big Bang. It is ironical that to describe the last moments of the Universe requires increasing understanding of the physical processes that governed its creation. Gravity, the framework of space and time, ruled at the outset and will step in when all else is gone.

In the first moments the heat produced a plasma of quarks, which rapidly clustered into the protons and neutrons that formed the nuclei of atoms when the temperature had cooled to a mere billion degrees. These conditions are still present in stars and have been accessible in experiments at particle accelerator laboratories for many years. These laboratory experiments have enabled us to study nuclear physics and learn about the stellar dynamics.

This symbiosis between nuclear physics in the lab and its natural realisation in stars was the first step in the unification of micro and macroscopic sciences that is currently in full swing.

Nature operates at the microscopic level and realises its aims by constructing macroscopic structures on which to play out its schemes. The human scale involves life operating at the level of cells, the chemistry of DNA working through on a large-scale host, the living being; we are complicated macromolecules. Stars are conglomerates of individual protons, neutrons and electrons working out their individual radioactive transmutations. The entire macro-universe is really a working through of beautiful fundamental interactions and laws at the atomic level, and even beyond. The deeper we probe, so the richer becomes our awareness of the possibilities for exploitation.

We know enough about the ways that stars develop to be sure that they will live only a fleeting existence in an everlasting universe. The Universe is some 20 billion years old so far. In another 20 billion years some new stars will have been born – many are being fused in Orion's belt right now – others will die. Stars like our Sun will use up their supply of hydrogen fuel in another 5 billion years.

Of local concern to us is that the Sun will then contract, heat up and expand to a red giant, evaporating the inner planets. The outgoing wind will vaporise the outer planets. The picture of the planetary nebula in the constellation Lyra shows what will remain of us.

It is to be hoped that by then we will have colonised a more favourable planetary system, fuelled by a main sequence star. But then the story will work through again there after another few billion years.

This nomadic enterprise can go on for perhaps 50 billion years as one after another 'new' main sequence stars come to their end. After 100 billion years, however, real change is beginning to show. Agoraphobics should quit now. The Universe has expanded so much that galaxies are exceedingly remote. Our nearest neighbours are barely visible in even the largest telescopes. The content of these galaxies is also changing. The stars are burning out and nothing remains to replace them. Every star has used up its fusion fuel and is collapsing under its own weight, a cold dark ball, a spent fire.

Neutron stars – the remnants of heavier suns – are fairly rare; most common are remnants of small stars – white dwarfs – balls of electrons and spent nuclei that are the size of our present Earth. There are black holes too; rarest of all but vastly exceeding those today. Advanced civilisations may play Russian roulette living on the edge of a black hole, tapping energy from it.

Some planetary systems survive. Charred remains of outer planets could survive red giant eruptions. It depends very much on what the planetary system consists of, and the size and distance of the planets from their sun.

Wait long enough and even improbable events are certain. Dead stars pass close enough to one another that they will fail to hold on to their planets. The planets will detach and drift off freely through the void. Even galaxies will get pulled apart as they collide. Again, this is exceedingly rare but we've all the time in the Universe.

Some 90–99 per cent of the stars will evaporate from the galaxies and leave an homogeneous spattering of bits and pieces around the cosmos. The remaining 1–10 per cent will tug inwards and collect in the galactic centres, forming gigantic black holes. This will be what things are like after 10^{19} years – 1,000 million times longer than from the Big Bang to the present. At this time-scale neutron stars will be cooling to only 100 degrees above absolute zero – −173 degrees Centigrade. Any planets still orbiting will radiate gravitational waves and decay after 10^{20} years.

The microwave background radiation is within one ten-thousandth of a billionth of a degree above zero. The Universe is a forbidding vastness of nothing. No galaxies, only stars. Lonely stars devoid of planets. The one-time planets are themselves dead carcases outcast by gravity, roaming at random.

Aimless dead wanderers through the void. What is there left for the Universe?

The decaying protons begin to be felt a little. Enough are decaying and producing radiation that they provide a trickle of 'warmth' – a fraction of a degree above ultimate cold – and can keep the ambient temperature at this meagre level for another 10^{30} years. So there is a lot of localised warmth for our descendants to exploit. The composition of the Universe is 90 per cent dead stars, 9 per cent black holes and 1 per cent atomic hydrogen and helium.

Proton decay is now going on in earnest causing whole bodies to die out. Soon all carbon-based life forms become extinct as the carbon decays: diamonds are not forever!

The time-scale to this epoch compared to the present is as the present age of the Universe compared to a single heartbeat.

By that time our descendants will have to change into other forms. The carbon and other matter that make up *Homo sapiens* will all be dying away as the protons decay. No longer trapped by

Nature's choice materials, we must exploit natural laws to manufacture new forms and utilise what remains. Discoveries in particle physics will be put to work; new opportunities may emerge for changing the very fabric of our Universe.

If we do not manage to make new forms of matter, the end-result of decaying protons will be a Universe containing electrons, positrons, neutrinos and photons. These will be our basic toolkit for survival. Freeman Dyson has some ingenious ideas on the continuation of life and information even in these extreme circumstances (see Chapter 13).

The distance between each electron and positron will be greater than the present diameter of the Milky Way. The Universe will be 100 million trillion (10^{20}) times larger than it is now. Even this is not the end.

According to the theoretical physicist Stephen Hawking, black holes are not entirely black but radiate energy and matter. This may provide material for us if we live near the edge of a black hole – if we throw our waste into the hole it will repay us energy and matter. So there may be bizarre chances for us yet. Ultimately even the largest supercluster black holes will have evaporated by Hawking's process and any new protons produced thereby will have decayed too.

Someone calculated that, if we survive all this, our descendants (if any) will meet descendants that have developed independently on a distant planet. This will be after 10^{300} years. How can we contemplate such time-spans? If creatures still exist then, will 'today' hold any meaning for them? Will Beethoven, Mozart and Bach be encoded for that distant future?

This is too remote to be of more than intellectual interest. But modern theories suggest hazards for the make-up of matter that could threaten us now or within a few years.

Stranger than Science Fiction

Nature always seeks the most stable configuration: water flows downhill; some atoms shed radioactivity to come to more stable forms. We are made of the most stable stuff around – atoms whose electrons encircle a positively charged nucleus made of neutrons and protons.

Although this is the most stable stuff around here it might not be the most stable stuff everywhere. Some theorists suspect that there may exist 'strange matter' that could initiate a collapse of ordinary matter were it to come into contact with us.

A powerful attractive force grips the neutrons and protons, tightly building the nuclei that are at the hearts of atoms. A single proton forms the nucleus of the simplest element, hydrogen, while as many as 250 neutrons and protons cluster to form the nuclei of the heaviest elements.

Protons carry positive charge and so mutually repel one another, destablising the nucleus. Neutrons, which as their name implies are neutral, do not experience this disruption and are easier to clump. Vast quantities of neutrons can clump together forming neutron stars, which are gigantic nuclei consisting of some 10^{57} neutrons. These are the two known forms of nuclear matter: collections of up to 250 or so neutrons and protons forming the elements of the periodic table and the mega clumps forming neutron stars. For decades we have believed that all matter in the Universe is of one or other of these types. Neutron stars might be exotic given the relatively quiescent conditions around this part of the galaxy, but they are made of the same stuff as we are at heart.

The 'strange matter' that the theorists are excited about is quite different. To understand the origin of the idea we must go back to 1947 when the first clues arrived here that there are more types of matter in the Universe than had been seen to that time on Earth.

Some 10 miles above our heads the outer atmosphere experiences a continuous bombardment by photons and subatomic particles. The photons cover the whole spectrum of electromagnetic radiation – from radio waves through visible light to gamma rays. Most of the other particles are atomic nuclei. These are produced in distant stars and, accelerated by the magnetic fields in space, they smash into the atmosphere with millions of times the energy released by radioactive sources.

After the end of the Second World War there was an urgent need to understand the make-up of atomic nuclei. The cosmic rays were ideal tools in this quest: they could shatter nuclei into fragments and even leave a permanent record in photographs of what was happening. Many scientists went up mountains or sent up film in balloons to record the extraterrestrials. When the cosmic rays passed through a small chamber filled with supersaturated air (called a cloud chamber for obvious reasons), they would leave a trail of drops similar to the trails left by modern high-flying aircraft. A camera could record these trails for posterity. A more direct method was to send photographic emulsion plates up in balloons and if a cosmic ray shattered a nucleus of one of the bromine atoms in the emulsion it literally took its own picture – revealed when the emulsion was developed. As a result of many hundreds of such pictures the make-up of matter was revealed as never before.

One day in 1947 Claud Butler and George Rochester from Manchester discovered a trail in a photograph unlike anything they had seen before. Today we know that this was the first recorded example of a 'strange particle', which has no analogue on Earth. Our atoms consist of electrons whirling around a nucleus containing neutrons and protons; strange particles are like neutrons and protons but somewhat heavier and intrinsically unstable.

This glimpse of forms of matter beyond our ken came just as the nuclear physicists were building the first of the great 'atom smashers', machines that accelerated subatomic particles like protons to near the speed of light and could simulate the effects of the cosmic rays to order. When these fast-moving particles smashed into atomic nuclei in their path, a maelstrom of particles poured out. These included strange particles, materialised from the energy of the collision (Einstein's ubiquitous $E=mc^2$ at work again).

The diffuse cosmic rays had produced only a handful of examples of strange particles, but the intense collisions in the accelerators produced them in thousands. Soon they were as familiar as the conventional particles such as the electrons and quarks that make up the matter around and within us.

Strange particles too are made from quarks, but they contain a variety of quark called a strange quark, not found in the protons and neutrons of our atomic nuclei. Why it should be that Nature isn't satisfied with the minimal diet that makes the stable matter that the Earth and the stars seem quite happy with, we simply do not yet know. Whatever the reason, there is no doubt that Nature can, and does, make use of strange quarks, building up strange particles and, theorists now conjecture, conglomerates called strange matter.

Individual strange particles are heavier than individual neutrons and protons and tend not to live very long, decaying into these lighter and more stable seeds of our atomic nuclei. However, the equations describing strange quarks suggest that conglomerates containing large numbers of them may paradoxically be lighter, and more stable, than iron – the most stable of the known atomic nuclei. This 'strange matter' could be the most stable form of bulk matter in the Universe and as such could cause our atomic nuclei to change into the strange form if they meet!

Having found the quarks, which are the seeds of nuclear particles and atomic nuclei, we understand better why it is that certain nuclei exist and others do not, and can contemplate the existence of as yet unseen forms of nuclei. It is possible that out in space there are bizarre stars where the neutrons have compressed into a huge plasma of quarks where individual neutrons are not identifiable. No one is sure yet whether this happens or not; there is still a lot going on 'out there' that we are ignorant about.

If strange matter does exist, there is a consensus that the electrons surrounding the strange nuclei will shield our own atomic nuclei from the strange form. Consequently it is possible that our atoms and the strange ones might get away with 'harmless' chemical interactions involving the peripheral electrons, and the cataclysm of our nuclei being changed by direct contact with the strange nuggets thereby avoided. Any that impinge on us from outer space will stop in the Earth's crust if they weigh less than one-billionth of a gram; greater than one-tenth of a gram and they will pass right through.

Some theories of the early Universe suggest that a large fraction of the mass of the Universe survived in nuggets of strange matter with sizes between breadcrumbs and oranges. This is concentrated nuclear matter, like that in neutron stars, and a thimbleful would weigh many tonnes. It is non-luminous, not the stuff of ordinary stars, and could be responsible for some of the dark missing mass of the Universe and the Milky Way. Some astrophysicists are suggesting that some examples of what we call neutron stars may in fact consist of strange quark matter. When such stars collide, or perhaps when supernovae erupt, pieces of strange matter may be ejected to pollute the Universe.

Strange matter hitting our outer atmosphere will shine like a shooting star but with some differences that can distinguish it from conventional meteors. The most noticeable is its speed – much faster than meteors.

Meteors are pieces of gravel that are orbiting the Sun, fellow-travellers in the solar system and moving at a similar speed to us – some 20 miles a second. Any moving faster than that would escape from the Sun's grip like a car failing to take a curve. If we hit one in head-on collision, like two trains travelling in opposite directions on the same track, the net closing speed would be 40 miles per second. The added pull of the Earth's gravity might speed the rock up to 50 miles a second but no more; this is the maximum closing speed for a conventional meteor. Strange matter, by contrast, comes from all over the galaxy and will be moving much faster than this. We are all orbiting the centre of the galaxy at about 150 miles per second. Strange matter that is part of the galaxy, but not trapped by the Sun, will also be moving at this sort of speed.

Arriving with such high energies they will reach the low-lying dense atmosphere before shining out. They glow at altitudes below 10 miles, unlike meteors which burn above 50 miles. Lower and faster than meteors, their angular velocity will be much greater.

Many of them will make landfall, similar to meteorites, one estimate being that as much as 1,000 tonnes could hit the Earth annually. If so much has landed we ought to be able to find some evidence for these 'nuclearites' as they are known to distinguish them from ordinary meteorites.

These nuggets of strange matter are much denser and more compact than ordinary matter. On passing through rocks they may etch out tracks as the nuclei in the rock recoil. There could be fossil traces of nuclearites that have impinged on the Earth over the ages.

A 1 gram nuclearite could leave a trail one-tenth of a millimetre in diameter. Nuclearites of the order of 1 tonne (still smaller than a pinhead!) would pass through the Earth in under a minute, causing epilinear earthquakes whose seismic signals will be very different from those of a normal point earthquake or underground thermonuclear explosion. Maybe there is evidence for large epilinear quakes with body magnitude greater than 5 awaiting 'discovery' in existing seismic data. Some theorists suggest that such an impact could occur once a year on average.

On passing through water the nuclearite will radiate light. Any hitting the ocean could momentarily brighten the lives of deep sea fishes but the chance of humans seeing any flashes in water is of a rather different kind. The underground experiments seeking evidence of decaying protons use huge swimming pools of water and phototubes to detect light flashes. If they are lucky they might chance upon a passing nuclearite in addition to (instead of?) the purpose that they were originally designed for! Particle physicists at accelerators are looking for evidence of small strange nuclei.

The theory of quark behaviour is very delicate and it is not certain whether it implies that strange matter exists or not. But it has at least made us aware that the observed stability of the familiar nuclei such as iron, carbon, oxygen and so forth need not imply that we are made of the most stable forms of matter. Radioactivity could change our atoms into the strange forms, but extremely slowly. The transition from the 56 neutrons and protons comprising iron (the most stable nucleus) into a nucleus of strange matter takes longer than the life of the Universe. So we are safe here on Earth at present. But it would be a bizarre joke if the most stable state for Nature was not realised somewhere. And if large amounts reach Earth someday, then . . . ?

12

Beyond the Fifth Dimension

North, east and upwards: we are trapped in a three-dimensional Universe which is evolving in the fourth dimension of time. We are so used to this that it is hard to conceive of things otherwise. What would a five- or six-dimensional Universe be like? Does it even make sense to contemplate?

One's immediate reaction is to say that it is clearly nonsense to imagine more dimensions. After all, where could you 'put' them! All possible directions are 'used up', already.

All of the ones that we can easily imagine are certainly used up, but that may be a statement about our lack of imagination rather than about the nature of the Universe. Suppose that we were 'flatlanders' – two-dimensional creatures living on a flat surface and only aware of the surface 'in' which we moved around all the time. That would be the extent of our Universe; the idea of 'up' would not be in the dictionary. Then someone asks the absurd question 'could there be a third dimension?'. With our greater awareness we can imagine 'up' and so may be surprised at the difficulty the flatlanders would have in making the intellectual leap out of the paper. In turn, *we* may be unaware of extra dimensions: call them 'beyond' or 'within'.

An ultimate theory of the universe ought to answer the question of what is 'magic' about the number of dimensions in which we find ourselves. It is one of the simplest questions to ask; but the simplest questions are often the most profound and difficult to answer. It is not at all clear how you would set about it scientifically.

However the current excitement in theoretical physics centres on a theory that makes profound statements about the fabric of the Universe. It goes under the codename 'superstrings' and implies that at the time of the Big Bang there were TEN dimensions. Six of these have become hidden to our gross senses but leave their mark in giving rise to electricity, nuclear radioactivity and related phenomena. The other remarkable consequences of the theory is that it might imply that there is an entire invisible universe operating right here inside the one that we are familiar with.

We cannot see this shadow universe but we can feel it. Its weight is tugging at us via gravity. It affects the galaxies and stars in their courses. Apart from the fact that it is there, we know next to nothing about it.

What impact this has for us is only now being worked out – the theory is very new, still only poorly understood and is being studied at universities and laboratories worldwide. One Nobel Laureate has described this as the greatest advance in theoretical physics since quantum mechanics or general relativity. This is praise indeed. These are the two great pillars of twentieth-century science; to compare the superstring theory with them implies that it may well be the holy grail of theoretical physics and, if so, its implications for a hidden universe should be given due weight.

The comparison with general relativity and quantum mechanics is interesting because superstrings subsumes both of these theories. Moreover it avoids an embarrassment that physicists have kept rather quiet about for many years: the foundations of science are flawed. General relativity and quantum mechanics apply to very different situations and have never been found wanting. But one can imagine circumstances where both of these theories have something to say – and it turns out that they are mutually contradictory. Superstrings shows the way out of this paradox and shows that it is in part due to our limited imagination – there are more dimensions in heaven and earth than we had dreamed of. Until we come to grips with this we are again like primitive societies who were all too aware of the unknown.

So first, let's see where science falls apart in four dimensions, meet the superstring theory that resolves the problem, and see what it implies.

Microscopic Quantum Theory

Our immediate senses are aware of structures larger than about one-tenth of a millimetre; simple microscopes extend that awareness to the scale of microbes. In the period up to the end of the last century 'classical physics' described the known phenomena in this macroscopic universe. But hints of profound novelties occurring at short distances were already reaching into our gross senses. Hot bodies emit electromagnetic radiation and the standard theory predicted a nonsense – that there was an infinite probability to radiate ultra-violet light. This does not happen; the paradox was called the 'ultra-violet catastrophe' and signalled a major failing of the existing world view.

The great German physicist Max Planck found the solution when he invented the quantum theory. This was an extension of classical ideas into the realm of microscopic distances. Matter is made of atoms, which are extended objects, about 10^{-10} m diameter, with a detailed inner structure. Classical physics is insufficient to describe phenomena at such distances, and the behaviour of microscopic atoms causes the radiation of ultra-violet light to behave rather differently than predicted in the classical theory. The infinite probability – the ultra-violet catastrophe – was avoided once scientists appreciated the essential role of the quantum theory, which implies that the laws have to be modified at short distances.

Quantum theory gave rise to a different world view. In quantum theory there is a fundamental 'uncertainty principle', in that you cannot measure both position and momentum or energy of a system to infinite precision. The more precisely a position measurement is made, the less precisely are the momentum and energy of that system knowable. This is imperceptible for macroscopic objects but more and more striking for phenomena at microscopic and subatomic length scales.

Einstein's famous equation, $E=mc^2$, implies that energy and mass are equivalent. Energy can, in a sense, coagulate and form particles of matter; conversely matter has the possibility to change form into radiant energy, as in certain nuclear reactions in the Sun for example. Now, according to quantum theory if you try to look at things at very fine distance resolution you find that the momentum and energy of the system under study fluctuate wildly – the more so the smaller the distance. The effect of the $E=mc^2$ is then that the energy fluctuations at short distances can be manifested as so-called 'virtual' particles and antiparticles (mirror images with the same amount of mass but opposite electrical charge to the particle); materialising out of the vacuum and surviving for a mere instant before they meet and annihilate.

At distances less than 10^{-13} m the fluctuations in energy are large enough that the lightest electrically charged particle – an electron – can be momentarily created along with its antimatter counterpart (the positron). As a result it is no longer possible to describe a system as containing a fixed number of particles: electrons and positrons are continuously materialising and disappearing on short time-scales. Nor is the vacuum a void – 'empty' space is a medium with an infinite number of particles and antiparticles seething within it.

All this is standard quantum theory as manifested at atomic and nuclear distance scales. It is the paradigm, essential in every theoretician's toolkit, and underwrites our theories of the fundamental forces acting on and within individual atoms. These are the electromagnetic force, which holds the electrons in the atomic periphery, the strong force, which binds the atomic nucleus, and the weak force, which is responsible for radioactivity ('beta decay') and the interactions of the ghostly neutral neutrinos. These three fundamental forces control all phenomena other than those due to the fourth great force – gravity. This force, known the longest, is in fact the least well understood. Superstring theory offers the promise of including gravity naturally in a unified theory of all forces and matter in the universe.

Gravity

Three centuries ago Isaac Newton gave the first quantitative description of one of Nature's fundamental forces with his celebrated theory of gravity. Although the force of gravity between atoms at earthly energies is feeble, *all* particles of matter attract one another gravitationally, with the result that the collective effects of many particles, such as those in the Earth, produce discernible effects, holding us on the ground, and controlling the motion of planets and galaxies.

When objects are moving very fast, the effects of gravity differ from those described by Newton's theory. One example of this is seen in the orbit of the flighty planet Mercury, whose point of closest approach to the Sun changes slightly from one orbit to the next. Einstein's theory of general relativity subsumes Newton's theory and to date agrees with all observations on gravitational phenomena.

In practice we apply the theory of gravity to eclipses, tides and the motion of satellites – namely bulk matter. What about gravity on the scale of individual atoms? The gravitational force is exceedingly feeble between individual atoms – it is overwhelmed by the electrical, magnetic and nuclear forces. When we are studying the behaviour of individual atoms and subatomic phenomena we use quantum mechanics but have no need of gravity or general relativity. Conversely, when we deal with large-scale structures that are interacting gravitationally we do not need quantum mechanics as this concentrates on the microscopic structure of matter. So in practice the two theories don't meet head on – one or other but not both are needed at any one time.

Einstein's general relativity was the first new theory of gravity since Isaac Newton's work in the seventeenth century. It includes Newton's laws and goes far beyond them, describing not just falling apples and the motion of planets and galaxies but also dealing with the evolution of the entire Universe. It makes profound statements about the relation between gravity and the nature of space and time. The best state of the art experiments cannot fault it.

The equations of general relativity show that it is *energy*, not simply mass, that gravitates. Light has energy and so gravity acts on it. The massive Sun can deflect passing light beams a little. If the Sun were much more massive it would deflect light beams a lot.

When enough mass is concentrated in a small region, the resulting gravitational forces may be so powerful that they entrap light and a black hole results.

The idea of a black hole in space is bizarre but not particularly mind stretching. It is so beloved of science fiction because of the extreme warping of space and time that occurs in its vicinity. The gravity is so powerful that space–time curl up on themselves: time, in a sense, stands still. It is here that the paradoxical conflict with quantum theory appears.

The Conflict

Quantum theory implies that energy fluctuates at extremely short distances or time-scales. We already commented that at 10^{-13} m electrons and positrons are continuously bubbling. At distances less than 10^{-35} m the energy fluctuations are so great that particles 10^{19} times as massive as a proton can form and annihilate. Such large masses concentrated in such minute distances are black holes. So quantum theory implies that very small black holes are coming and going. At very short distances, or on very short time intervals, time is standing still. Gravity is distorting the environment so much that our whole notion of space and time breaks down.

The problem is that our quantum theories of force fields ('quantum field theories') are built on the assumption that it makes sense to talk about space and time at all length scales and over as brief a time interval as you choose. Time is regarded as continuous, ever-flowing and not standing still or jerking about. Thus, general relativity (black holes) and quantum theory are fine if kept apart but must be modified in an ultimate theory.

If you ignore these conflicts and attempt to calculate numbers anyway you find nonsensical results. Quantities that should be finite in reality turn out to be infinite in the theories. For example, the gravitational force between two electrons, 10^{-22} cm apart, is predicted to be infinite.

It is not new for infinities to appear as the answers in quantum theory calculations. They occur in quantum electrodynamics (the quantum theory of the electromagnetic force) all the time but are 'harmless' in that they can be removed by a well-defined mathematical technique called 'renormalisation'. In effect the infinities disappear by a redefinition of what we mean by mass and electrical charge of the particles such as the electron. You only need to do this once and it works all the time. That this can be done consistently is profound and also practical; it provides a quantum theory of electromagnetism consistent with relativity and common sense.

Unfortunately this doesn't work in the case of general relativity. Einstein's theory implies that gravity's strength gets more and more powerful the more energy that the gravitating particles have. This turns out to undermine the theory – infinities pop up that can no longer be absorbed and we seem to be stuck with them. This is clearly a nonsensical situation and something has to be done.

For a long time people have tried one idea after another and failed. Nothing in the conventional wisdom seemed to do the job satisfactorily. Here we have an unwanted and paradoxical infinity reminiscent of the ultra-violet catastrophe that heralded the birth of quantum theory. Some new ingredient, or new theory, is called for, subsuming general relativity and quantum theory. This is what superstring theory appears to have achieved.

Just as the black body radiation paradox was the first clue to richer structures occurring at distances below our gross perception, so is this new paradox. In the case of the infinite black body radiation, atomic granularity was the answer and quantum mechanics the new dynamics. Quantum mechanics goes over into the conventional mechanics at large distances where the graininess is hidden. However, the discrete spectral lines emerging from atoms – the fact that different elements radiate characteristic spectra of colours, such as the yellow or blue of sodium or mercury street lamps – are a reminder of what lies beneath.

The paradoxes that confront general relativity and quantum mechanics are clues that there is a graininess at extremely short distances, far smaller even than the scale of the atomic nucleus. According to the new superstring theory, Nature has a complicated and detailed structure on scales millions of billions times smaller than known atomic particles such as electrons and protons. What we thought of previously as points are now regarded as extended structures vibrating like violin strings. (That is the 'string' part of superstrings. The 'super' refers to a particular property of the maths that isn't relevant to our present story.) This graininess contains six hidden dimensions that extend less than one-billionth of one-billionth part of the size of a proton.

A large-scale remnant of this deep richness is the possible existence of a dark universe. The hidden dimensions also leave an impact. In our four-dimensional perception they manifest themselves in electrical and nuclear forces.

The Fifth Dimension

The idea of a fifth dimension goes back more than 50 years to the work of Theodor Kaluza and Oscar Klein. Einstein's theory of gravity – general relativity – treats time on the same footing as space, so it is a theory of 'space–time'. Our Universe exists in 'space–time' – a total of four dimensions.

Einstein's theory is constructed in such a way that you can write its equations for mythical universes with more than four dimensions. This is a mathematical game that Nature appears not to care about since in practice we live in only four dimensions. Kaluza and Klein wrote out Einstein's theory in five dimensions and then examined what happened when the fifth dimension was siphoned out.

If you just throw it away then you are back to standard four-dimensional gravity and have achieved nothing more than a waste of paper. Somehow you have to keep the fifth dimension and yet hide it because it doesn't manifest itself in any obvious way to our gross senses. The way they did this was to suppose that the fifth dimension is curled up, compacted, existing only over very short distances (see Figure 12.1). What causes this is not known and is a question for the future. For now, just impose it on the equations and see what happens.

Well, Einstein's gravity emerges from the four dimensions that are left untouched, which is no surprise. The interesting thing is – what happens to the gravity in the curled-up dimension?

The astonishing result is that the equation describing the

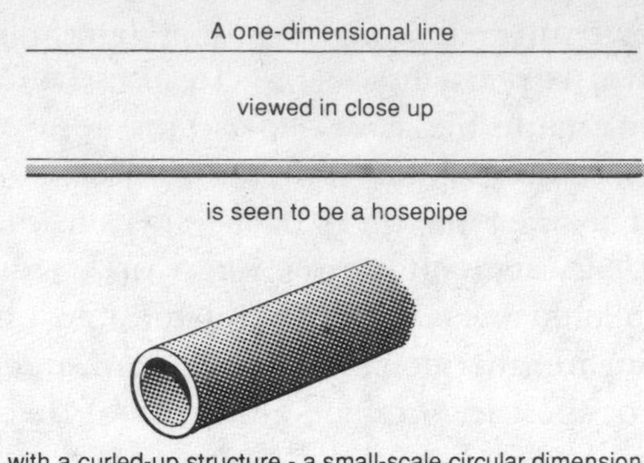

12.1 Curled dimensions

gravitational forces in the fifth dimension is one that Kuluza and Klein had seen before and that is familiar to every student of physics. It is the equation that James Clerk Maxwell discovered in 1865 and that describes the *electromagnetic* force. What Kaluza and Klein's work had led to was the realisation that what we call electric and magnetic forces are in fact gravity – in the fifth dimension. When you play with magnets, or turn the starter on your car engine, you are in contact with the fifth dimension.

Today we know of other forces that act in and around the atomic nucleus. The weak force that gives rise to radioactivity is one. We know that it is intimately related to the electromagnetic force and so we expect that it, too, is gravity in higher dimensions. The modern grand unified theories imply that the weak and also the strong forces binding quarks and atomic nuclei are subtle manifestations of the electromagnetic force. The notion of higher dimensions asserts that they are all related to gravity.

The mathematics describing the weak and strong forces is more involved than that for the electromagnetic, and a total of five more dimensions is needed to bring them all together. So if the electromagnetic force is gravity in the fifth dimension, the weak and strong forces are gravity in the sixth–tenth dimensions.

There are several indirect hints that these ideas are right, that there really are more than the three space dimensions and one time dimension that we are familiar with. The theorists are now grappling with the problem of why three space dimensions grew into the macroscopic Universe while the others curled up and left their mark in other ways, such as the electrical and nuclear forces. In the first instant of the Big Bang, space and time were undergoing contortions that our mathematics is not yet able to describe. But the majority of mathematical physicists specialising in particle physics, cosmology and astrophysics are working on this problem or on closely related ones. No one has all the answers yet but pieces of mathematics are emerging from the study of superstrings and from the study of the early Universe in general that have suggestive implications.

One of these is that space and time, as we know them today, might be unstable. This is a terrifying thought but not necessarily absurd. After all, we don't know why there are the number of dimensions there are; we have clues that they are in a sense mere 'remnants' from ten, so why should they be permanent and unchanging? How guaranteed is the fabric of the Universe within which we find ourselves?

The Collapse of Space and Time

As I commented in the previous chapter, Nature always seeks the most stable configuration – the state of lowest energy. There we met the possibility of strange matter, more stable than our own, that could seed the collapse of our matter should they meet. A more dramatic possibility that has emerged from studies of the Big Bang is that the Universe as a whole might be inherently unstable. I don't mean here that its matter is eroding away on an indefinite time-scale too remote to concern us; instead I'm concerned at the possibility that the very fabric of the Universe could suddenly change somewhere *now* and spread out like a cancer at near to the speed of light, devouring everything. All the bits and pieces of matter at the deepest level remain the same but reconfigure – change 'phase' is the technical description.

The notion of matter changing phase is very familiar. Ice, liquid and steam are different phases of water; the molecules of H_2O are the same in each case but bind collectively in different ways. Which configuration they choose depends on external conditions such as temperature and pressure. You can superheat it or supercool it. For example, the water in your car radiator can heat above 100 degrees Centigrade when stuck in a traffic jam. The pressure prevents it boiling, so long as the radiator cap is efficient. Unwise motorists who start to remove the cap can receive a blast of scalding steam as a result; the superheated water has boiled as the pressure drops. Similarly water can be supercooled below freezing point, remaining liquid so long as external conditions don't change, but if they do then the water suddenly freezes. In its supercooled state we say that it is metastable, while the frozen form is stable.

A popular idea among some physicists at present is that the Universe is in a metastable state – that the cooling from the hot Big Bang left us with a supercooled rather than a frozen Universe. Stars, planets and human beings are the natural order of things given the building blocks that Nature has to play with and the way the laws of our Universe prefer them to be combined. But if the Universe is only supercooled and should freeze in the future the basic building blocks could be combined in a more energy-efficient way. You, I and everything that we've ever known would go and a new order prevail.

Already, around the cosmos, we can see examples of how basic particles form different microstructures and hence different forms of bulk matter. On Earth today electrons, protons and neutrons are bound in atoms, whereas in the Sun these same particles roam free in a plasma. In a sense matter is in a different phase here than in the hot stars; it consists of the same particles but bound tightly like ice rather than more freely as in water. The paradigm among theorists is that the Universe started very hot, hotter than any star today. In the Big Bang even the protons and neutrons melted; in those early moments their quarks were freed. Earlier still, when conditions were even more extreme, it is possible that the Universe was in some other phase; its cooling into the quarks and eventually nuclear particles could have left it in the true frozen state of lowest energy or maybe only in a supercooled state where it has remained – so far.

It is hard to imagine what a more 'frozen' form of these particles would be like as we have no intuition to guide us. All we can say is that we have no guarantee that we are the favoured end-product. If we are not, we could still be lucky – the Universe could be stuck by chance in its present phase and have no manifest way to reach its most favoured configuration.

In analogy, we are on the first floor of some cosmic cathedral, which is fine so long as there isn't a trapdoor for us to fall through, or at least, if there is, that no one opens it by mistake.

The problem is that quantum theory can 'open the door'. Tunnelling through barriers (the floor) to fall from a high-energy state to a lower one is at the root of many nuclear decay processes. So wait long enough and somewhere Nature will tunnel through from our metastable to the true ground-floor configuration. And the Universe has been around for over 10 billion years – long enough that somewhere, sometime, such a change may have or even now be taking place. This could happen at anytime, anywhere.

If a bubble of 'frozen universe' forms, it might die out or grow uncontrollably; it depends how much lower in energy this desired state is relative to the metastable one – how far above the ground level our present floor is. A growing bubble would expand at nearly the speed of light with enormous release of energy. Everything would make a big *vvarroomph*; the roar from strange matter would be a whisper by comparison.

In 1983 the distinguished astrophysicists Martin Rees and Piet Hut introduced a new feature into this debate by asking in the journal *Nature* whether modern technology might open the trapdoor inadvertently. Could a new accelerator of subatomic particles produce such a concentration of energy in one spot that it flip that part of the Universe and, like a cancer, ultimately the whole cosmos through the trapdoor – or to be more formal: 'make a spontaneous transition via quantum mechanical tunnelling'. As new particle accelerators reach into regions of energy not before probed on Earth, this awesome thought gives pause.

After some study, Hut and Rees concluded that, thankfully, we probably don't need to worry. Although it is the first time that the new accelerators will have produced such collisions on Earth, Nature is doing them all the time. Cosmic rays consist of nuclear particles at huge energies, far greater than yet produced on Earth. They are very scarce in space but even so several hundred thousand extreme collisions have occurred in the Universe in history and the Universe still survives. So Hut and Rees concluded that no particle accelerator in the foreseeable future will pose any threat to the fabric of our Universe.

But what if there are advanced beings in Andromeda who are building accelerators billions of times in excess of anything that we can even dream of? This game is the ultimate pollution – a bubble spreads and devours everything. Thankfully none is about to devour us; it travels close to but less than the speed of light, so we will have some advance warning of its arrival. I'm unsure what we could do about it if we saw one coming, but I take refuge in the hope that any people far ahead of us are advanced enough to have thought this through already and either not built the machines or surrounded them with the ultimate radiation shield!

Cannibal Universes?

Is it possible to create a universe in the laboratory? This sounds like the ultimate science fiction dream, but by 1987 understanding of the origins of our Universe had reached such an advanced state that the eminent astrophysicists Ed Farhi and Alan Guth from MIT debated the idea in the journal *Physics Letters*. They concluded, 'As you may have imagined it is rather difficult' but, within quantum theory, perhaps not impossible in principle. It is impossible in practice only given present technology and therefore, as I cautioned in the previous example, it may be accessible to advanced creatures elsewhere.

Until very recently we thought that the Universe is permanent and omnipotent. A few years ago we thought we had figured out Genesis; we argued about the long-term future, such as whether the Universe would expand and cool or suffer a heat death in collapse, but we agreed on one thing – ours is the only universe around. Catastrophic things might happen to the matter within our Universe, but the Universe itself – the space and time within which the drama plays out – that would go on. Now that dogma is being questioned. Andrei Linde, a distinguished Soviet theorist, even proposes that our Universe actually consists of innumerable separate mini-universes, whose laws may differ radically from the one in which we happen to exist.

Suddenly the Universe seems much less stable, less certain than it did.

Once we thought that the Earth was the centre of everything, then the Sun and finally we have become used to the idea that we occupy an insignificant outpost of an unremarkable galaxy that is but one of billions. Having had our egocentricity so rudely displaced one might have thought that that was the end of it. But no; now we are even questioning whether our Universe is the one and only – are we living in a polyverse?

The idea that we could create a new universe in the laboratory comes from the current widely held theory that in the first fraction of a second our Universe inflated drastically before settling down to the more leisurely expansion of the last 20 billion years. This 'inflationary universe' theory explains some puzzles that had plagued earlier models of the Big Bang and makes testable predictions for the present state of the Universe which have been successful so far. According to the inflationary theory, the Universe originally had a mass of under 10 kilograms in a size one-billionth of an atomic nucleus. Ten kilograms isn't much; you could check it onto an airliner without paying excess baggage charges. This is enough to start a universe, so how did it grow?

According to the theory, the interior of this region was metastable and, upon going through a phase change into the stable state that we are now in, it gained energy courtesy of a peculiar quirk of the quantum theory. The effect was like antigravity – a vast repulsion and growth of net energy until it settled in the present phase, its residual expansion being the fossil remnant of that trauma. Enough energy was generated to boost that initial 10 kilograms into the entire Universe that we see today.

So, if a universe can emerge from no more than a suitcase, might we make a cauldron in a finite region of space and arrange for it to be in the metastable state? And then, whoosh – this little region bursts out as a universe with a future evolution similar to that of our own, the whole thing starting off in the middle of the luggage check-in counter?

If someone made a new universe, what would become of our present one? Are we at risk from the experiments of some DIY enthusiast in Ursa Major? After all, can more than one universe exist? And, if it is possible to create a new one, then surely it is possible to destroy our own?

The theory that suggests this is science fact also seems to imply that we are safe if new universes are erupting. They make their own space and time without spilling into ours, so they won't destroy us. As far as we are aware, the wall of the bubble would be like the surface of a black hole.

It is hard to visualise this idea of a universe appearing apparently out of nowhere and yet being somehow disjoint from the space and time in which we exist. This is all happening in dimensions beyond our immediate experience so, as we did earlier, it may help to imagine the universe of the 'flatlander' and how this spontaneous eruption appears there.

The flatlander's universe is the surface of a sphere that is so huge that it appears completely flat, with no curvature at all. The formation of the new universe is like an aneurism on the surface where space and time suddenly bulge out and then separate from their parent into a new universe. We can visualise this, but flatlanders can say only that the new universe is 'somewhere else', as they are limited to their flat comprehension.

Someone living inside the bulge might experience conditions similar to a Big Bang. To people living elsewhere the bulge would appear as a black hole. After separating, the black hole would have appeared to have evaporated, leaving no trace of its formation. Like the flatlanders, we could say only that its bulge existed 'somewhere else'. We would say that the new universe existed in a totally separate space and time. Once gone, we and it can never communicate again. You can't get from there to here, nor can it invade our space. It could even be that our Universe split off from a host courtesy of some do-it-yourself experiment 20 billion years ago.

So the bottom line at present is that science admits the

possibility that new universes may erupt and that our present one could collapse. The former will not harm us and the latter is exceedingly improbable. In which case we are safe from these ultimate apocalypses – at least according to our current understanding. But five years ago no one would have seriously posed these ideas; today they are being worked through and the final outcome is anybody's guess.

13
Time Is Running Out

When the Sun Becomes a Red Giant

It is dawn in the Garden of Eden. The Earth spins us from the shadow of night into the bright light of day. Within a few hours the heat in the desert will be so intense that even the lizards will be seeking shade. The houses on the twentieth-century hills of Moab all have solar collectors on their roofs. They are accumulating the heat from a nuclear reactor 100 million miles away.

Solar energy – non-polluting, free of charge and permanent – utopia. But nothing comes free; there is a price to pay, even though the debt will not be called in for a long time yet. The desert heat is the end-result of the Sun consuming 600 million tonnes of hydrogen every second. So although it is free as far as we are concerned, it is a very inefficient way of providing us with fuel. Our tiny planet intercepts only one part in 1 billion of the whole; the rest goes out into space.

Several billion tonnes of hydrogen have been used up since you started reading this sentence and by this time tomorrow 100 million million tonnes will have gone for good. This is a lot, but is such a trifling fraction of the whole Sun that we don't notice it day to day, or even over a lifetime. During the entire million years of human existence the relative amounts of hydrogen fuel and its helium end-product in the Sun have changed by less than one part in 1,000. However, slowly but surely the Sun is changing, dying; wait long enough – about 5 billion years – and it will be all used up. What then?

When all of the hydrogen has gone the centre of the Sun will no longer be able to resist the weight of the outer regions pressing inwards. The core will begin to collapse and in doing so its gravitational energy will turn to heat in a re-enactment of its birth from a collapsing dust cloud.

As the core collapses, its heat throws the outer regions upwards, expanding and cooling the Sun's surface. The Sun, which has risen golden over the eastern horizon for 10 billion years, begins to grow bigger and redder. Imperceptibly at first, but then gathering pace as the outer surface approaches Earth, the Sun occupies more and more of the sky. The polar ice caps melt and flood low-lying areas, altering for ever the terrestrial geography. The rising temperature evaporates the oceans, making the entire globe like a tropical rain forest covered with a perpetual cloud. Viewed from outside we would appear shrouded much as Venus looks to us today. Intense redness pervades the cloudscape until the heat boils the oceans entirely, the clouds and atmosphere evaporate into space and the barren planet is defenceless in face of the red giant that fills the whole of the daytime sky. Dante's inferno has finally brought hell on Earth.

There is no hope for Mercury and Venus, which have been consumed within the swollen Sun. If Earth lies outside it will be scorched, lifeless and desolate. Our descendants may survive underground but that merely postpones the end.

The helium that is the end-product of the hydrogen fusion can itself fuse to build up heavier elements such as oxygen, nitrogen and carbon. This is what is going on at the heart of the red giant. In doing so the Sun heats up one last time, suddenly swelling out far beyond the Earth, vaporising it and Mars too as it expands to almost the orbit of Jupiter. The Sun's atmosphere spews into space, departing the solar system; viewers from other parts of the Milky Way will see a nebulous gas, the remnants of a one-time life-giving star. Examples of such nebulae are visible to us in our own night sky; we can only guess what life forms they once may have supported.

This Apocalypse is certain, guaranteed, unavoidable, unless we learn how to intervene. It will be the end of this jewel, our home and repository of our entire accumulated culture. It signals a pivotal moment in the future history of the human species. What hope do we have? Humans have existed for a mere trifle in the Sun's time-span. In a few thousand years we have progressed from fire to nuclear power, from the Stone Age to silicon chips, from scavenging for food to chemically synthesised vitamin tablets and meal pills for space travellers. Imagine this time-span passing once more, and then again 1 million times; on that time-scale we might be able to control the evolution of the Sun and thereby postpone Apocalypse. At the very least we will surely be able to colonise other star systems.

First we must plan for survival on planet Earth so long as it exists. We will have to decide what to do if we are threatened by impacts from comets, asteroids or the many hazards that can be expected during another 250 orbits around the Milky Way. This will take us through dust clouds where new stars are forming from the debris of old ones, where some day we will find ourselves near to a supernova or have our orbit disturbed by encroaching too near some other star. Any number of things could happen before the Sun becomes a red giant.

It should be possible to prepare for such things, to leave the planet and colonise space. Flying the Atlantic would have been regarded as an impossible dream when Columbus first sailed, yet today it is part of a common vacation for millions. Colonising space in another few years seems to me to be a natural extrapolation from today. And we know that we must do so to survive.

If we colonise new worlds we will then be faced with life after our ancestral Sun has died. We will see whole galaxies of stars die out as the entire Universe moves inexorably to its own end. For the Universe is a living thing, evolving as we do, eventually to die. On that time-scale we too will have evolved into new forms. Present life has emerged from simple molecules in under 4 billion years, so after a further 5 billion our remote descendants will bear as little resemblance to us as we do the first amoeba. Indeed, the need to change our form, if not just get off the Earth, becomes increasingly important as time progresses because if the Universe survives the sudden extinctions that we have mused over, it will eventually collapse in heat or expand into ultimate cold. Flesh and bone will not survive either of these; transmogrified humans might.

Princeton University theorist Freeman Dyson has thought about this and I will describe some of his ideas later in this chapter.

First things first: what we can do about more immediate problems. Then we will take a look at our prospects for surviving into the deep future.

Prepare for Collision

In the short term we may have to face the prospect of being hit by an asteroid. We have been struck by moderate-sized objects in the past, even in this century. The future is likely to be no less risky. Sooner or later we are going to find ourselves on collision course with a real monster. What can we do about it?

Remarkably little research has been done into this. Perhaps it is difficult to put one's mind to work on something whose need is not immediately urgent. There are, after all, any number of problems to deal with that already threaten our species, most of them self-inflicted. If we are unable to come to terms with self-imposed Armageddon, what hope do we have of making a collective effort against natural extinction?

So, are we to give up all hope? Is the human existence to be a great joke of Nature – a brief spell of light and intelligent awareness of the cosmos between two great expanses of darkness? The politicians of the world have their contribution to make to this ultimate question. The scientists should be also addressing the question of natural catastrophe. Joseph Smith of the University of Chicago proposes that the scientific community initiate a 10-year study on the prevention of impacts with a view to examining the costs and practicalities of the task.

First of all we need to know the number and orbits of asteroids much better and detect all large ones that have paths intersecting our own. Regular satellite launches should occur carrying equipment capable of detecting large Apollo objects (Earth-crossing asteroids) and new comets. The strategy would be to scan rapidly those orbits where asteroids are most densely populated, followed by a routine calculation of their orbits. You then have to observe them again later to check that the orbits have been correctly determined. Soon you will be able to say with a good degree of precision where they are going to be and when.

Later, as experience grows, this could be extended to smaller bodies (even 10 metres across can pack a fair thump after all). It is hard to detect small objects far away in space and so an array of telescopes will be needed to cover the sky.

Forming a timetable of asteroids is well within the capabilities of present technology. It merely needs the will. The *real* concern, however, is that it first requires the awakening of consciousness. I hope this book has made clear that this *is* so. Moreover, it is a worthwhile scientific endeavour in its own right and will stimulate development of instrumentation that will find applications far beyond the original aim. Billions of dollars were spent sending a few men to the Moon; the asteroid watch is small by comparison.

This is the first step – know your enemy. Next we have to do something about it.

A world that is becoming enamoured of the 'Star Wars' defence initiative is rapidly being carried away with the omnipotence of technology. The problem of dealing with an approaching lump of rock is 'simple'. Just send up enough rockets loaded with thermonuclear devices and blast the rock to pieces. As we saw in Chapter 4, Hollywood chose this solution in one of its disaster movies. This is as far-fetched as the belief that we can build a shield of lasers and weapons above our heads and rely on it working 100 per cent efficiently on the day it is needed.

The Faustian proposition of sending up thermonuclear devices may be avoidable if we detect an asteroid on a looping orbit that is slowly closing in on our own on successive circuits. We would have many years to settle on a course of action – a small deflection could be enough to make us safe.

Another area of ignorance concerns the make-up of asteroids and comets. The Giotto mission to Comet Halley has shown us more of comets than we ever knew. They are tenuous and less likely to be of concern than asteroids. Although bits of asteroids have landed, our knowledge of asteroids in pristine condition is minimal.

We need a systematic mission of flybys to both asteroids and comets. We have sent robots to Mars, so we can land on an asteroid. This is essential if we are to determine their properties. For example, we need to know if they are solid and hard or crumbly. Is there a lot of flaky rubble on the surface? Are they fragile and susceptible to artificial break-up? Is it easy to fly around an asteroid or is there a cloud of debris that would create a hazard for a team of astronauts intending to deal with it?

This is the most likely natural global catastrophe, and we can begin to tackle it with relatively unsophisticated means. We need not sit around passively awaiting extinction if we have prospects of developing the technical means to deal with the problem. Some of the plans that attempt to survive nuclear holocaust may have lessons for a global catastrophe of the sort I have been discussing. Some of the effects of a huge collision will be like those from global nuclear war. The nuclear winter, caused by smoke from burning cities, may well have its analogue in the dust cloud sent up by an asteroid disrupting the atmosphere. It takes more than two years for the atmosphere to clear itself of aerosol spray particles, so this is the period that we need to contemplate surviving. The Swiss have a national plan which is a good example of where to start. It is compulsory to have underground shelters for all individuals permanently stockpiled with two years' supply of food and other essentials.

However, nuclear war is a more likely bet, in the short term, than an asteroid impact; and, as very few are seriously planning to outlive nuclear war, there is not much hope of these ideas being put into action. I am merely reporting the facts in the hope that people will eventually become concerned enough to act on them.

New Worlds

We may learn how to deal with asteroids, comets and other natural catastrophes but sooner or later we will have to quit the planet. The Sun will eventually become a red giant and we must take ourselves beyond its range of destruction. If we continue to live near the swollen Sun we will have to adapt our energy supplies to the new conditions. We may have to go and colonise another star system, or even leave the galaxy as the stars die out or collapse inwards to form a great black hole. The chances are that we will have destroyed ourselves in thermonuclear war or industrial pollution before we have the chance to see what are the limits of technological achievement.

None the less, we have to make a start. It is unrealistic to plan far ahead but the first step was already taken when Neil Armstrong stood on the Moon. Many people still alive today were born before commercial aircraft existed; now they see us venture into space. Within a decade of marvelling at Yuri Gagarin and John Glenn in their pioneering orbits of the planet, we had reached the point where trips to the Moon ceased to be media events. By 1986 private citizens were being seated on the space shuttle as if the new frontier were already conquered.

The Challenger disaster made many people question science and technology, as if humans are preordained to remain earth-bound and are tempting fate to try to fly. Yet in the same month we saw the success of the Voyager mission, which flew past Uranus on cue, sending back pictures of an outpost of the solar system, something that would have been thought impossible 20 years ago. And for the first time international collaboration sent a team of rockets out to meet Halley's comet.

The technology for space exploration and development certainly exists already. Is the best motivation for future research really to put lasers and mirrors in space for military purposes? So far refugees from war have been able to flee to other countries. Where will a refugee from thermonuclear world war be able to hide? When car exhausts have polluted the air, trees been destroyed for cigarette papers (and books!) and life in the industrial environment become utterly impossible, where do we go? At some point, and very soon, we have to start thinking seriously about this.

The science fiction writers know the answer: they have us populating other planets, asteroids and purpose-built space stations. The idea of building space colonies has been looked into seriously by some scientists, notably Gerard O'Neill of Princeton. He started the project in 1969 as a stimulus to a physics course that he was teaching at the university. Gradually it became clear to him that the idea was feasible and that the required technology is for the most part already available.

He has written about his ideas in some detail in the magazine *Physics Today* (September 1974) and in *The High Frontier* (Jonathan Cape, 1976) so I will review his ideas only briefly.

People need air and water, land, gravity and energy to keep going. We could have all of these if we lived inside huge cylinders, several miles long and a few miles diameter. The cylinders rotate, so we would be pressed to the ground as in a centrifuge – though we would experience it like gravity. Solar radiation provides the power, enough to satisfy our needs ten times over at present consumption. In cylinders of this girth it is possible to build towns, fields, forests, even mountains to give an Earthly perspective and minimise psychological 'home sickness'.

A major barrier to space travel is getting off the ground. Compare the huge rockets that sent the Apollo astronauts on their way with the small craft that later lifted them from the surface of the Moon. The Moon's gravitation is so weak that anyone can be a high jump champion there. It is an ideal launchpad for space ventures. Moreover, the Moon is full of minerals, which could be mined and lifted into space to build the colony. In the longer term the asteroids could also be mined.

Using materials from outer space is the big jump. So far, space missions have relied totally on supplies brought from home, like early voyagers on the seas. Then they arrived in new lands, discovered new things and exploited the discoveries. So it can be in space. Mining in the solar system will be the great liberator.

So, how to get started?

It should already be feasible to build a prototype of 100 metres radius and half a mile long. This could support a population of a few thousand people! Rotating every 20 seconds it would reproduce the Earth's gravity. Present technology could accommodate a cylinder of 4 miles radius and 20 miles long, which could house 1–20 million people as the upper ecological limit. The atmosphere would be like the Earth's at a height of 2 miles, so it takes a little acclimatisation but is quite healthy as mountaineers will testify. The clouds will be similar to those on a summer's day.

In fact if you have two parallel cylinders rotating in opposite senses they can maintain their orientation as they orbit the Sun. The surface of each cylinder is divided into land areas and windows. As the axis of the cylinder points towards the Sun, you can't directly see sunlight through the windows. There is no external atmosphere to scatter light and make blue sky so you need to have mirrors outside reflecting the sunlight in through the windows.

The mirrors rotate with the cylinders. By opening or closing covers on the mirrors you can increase or decrease the amount of light. So night and day can be controlled in length to set the average temperature within the cylinder and reproduce the seasons.

Solar power stations are placed at the sunny end of the cylinder, so as to receive the maximum unshielded radiation. These will produce enough power for the inhabited cylinders and also the agricultural cylinders that accompany them.

Agricultural areas encircle the main cylinder. Their climates can be distributed through the seasons so that there is always a fresh supply of different crops. The design of these cylinders can be more functional; there is no need to include 'psychological' aids such as mock Earth features.

The early chapters in this book dealt with the hazards of collisions. The space colonies will be more vulnerable as they have no external air shield to burn up the stones; on the other hand, they are a much smaller target than the Earth. Most meteors are cometary, more snowballs than rock. The chance of being hit by a 1 ton lump is one in 1 million years. A piece of 100 grams will hit once in three years on average but well-designed craft should be able to deal with that.

Travelling inside a cylinder could be by bicycle or by electrically powered pollution-free vehicles; with an extent of a few miles only there would be no need for more than this. Travel between cylinders involves power-free craft using the natural spin of the surface to speed them on their way. The surface is moving at nearly 400 miles an hour. Unlock a craft on the outer edge and it will fly off tangentially across the intervening space at the speed of a jetliner. By swinging under the other cylinder it will be able to dock at zero

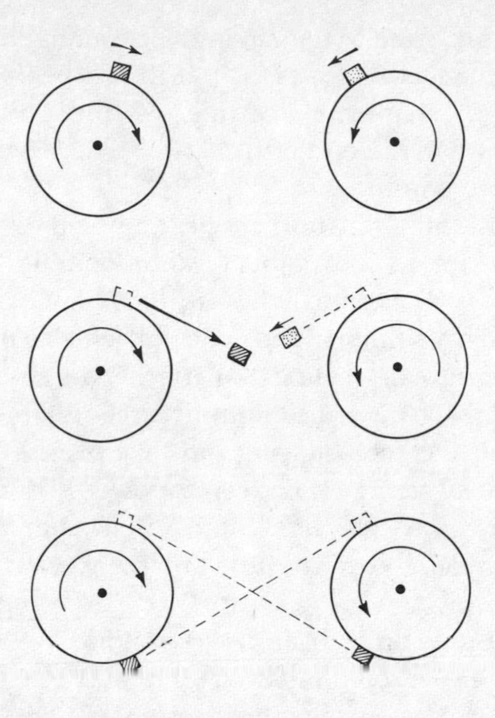

13.1 Travelling between two rotating cylinders The cylinder on the left is rotating clockwise. The vehicle (the box on the top) departs at zero relative velocity to the surface, which is very fast relative to the centre of the cylinders. It arrives at the other cylinder, which is rotating anticlockwise, and is captured, again at zero relative velocity.

relative velocity – like the passage of the ball between two lacrosse sticks (see Figure 13.1).

A series of cylinders 100 miles apart could support a total population larger than that of the present Earth. Different cultures could occupy different cylinders if desired – the possibilities for social experimentation are beyond present evaluation. O'Neill hopes that the 'need' for conflict and war would be reduced. The pressure for land and better lifestyle would be less than on Earth certainly, though I somehow doubt that human society is that straightforward. There may be insuperable unexpected psychological problems. For example, in the 1950s high-rise tower blocks were thought to be a utopian solution to slum clearance. The problems that they generated were unforeseen; many occupants would have preferred to remain in the unsanitary conditions from which they had been 'released'. Will we be rediscovering the isolation of the high-rise apartments? We understand human nature poorly enough on Earth that it is hard to extrapolate to an entirely new environment.

Cable cars could transport people among the cylinders in a few hours. There are no constraints of aerodynamics and so the craft could be roomy and comfortable, quite unlike the cramped conditions on modern airliners. The external vehicles that are whipped from the surface of one cylinder to another are so simple to operate that family trips to distant communities should be feasible. Polluting sports such as motor racing can take place in special venues. New sports may emerge such as manpowered flight. The effective gravity inside the cylinders comes from the rate of spin and decreases as you go up from the surface towards the axis of the cylinder (at which point it has vanished). So you get lighter as you climb the mountains and can fly at altitude.

In principle we could build the craft and place them where we like. In practice it is obvious that they should be within easy reach of Earth and the Moon. This is also at our 'natural' distance from the Sun. We must avoid being eclipsed by these natural bodies too often as our power supplies depend on sunlight. The craft must be stable against displacements in three dimensions. The Sun and Moon raise significant tides on Earth; bring the Earth into the system as well and there will be a complicated mix of forces on the spacecraft. The solution to the equations was recently worked out and it is possible to park the craft at various special locations in our vicinity where they will orbit the Sun with us while making excursions of some tens of thousands of miles. In the long term it would be possible in principle to put several colonies in such orbits.

So let's start building.

If we get lunar material we can exploit the Moon's low gravity to get things easily up to the construction site. Indeed, components could be pre-assembled on the Moon. There are plentiful oxides on the lunar surface which can be used to provide oxygen for water.

The total mass of a prototype station is around 500,000 tonnes.

We need aluminium, titanium, silicon and oxygen; 98 per cent of this can be obtained from the Moon. Add nitrates and trace elements to lunar soil and you have a viable agricultural base. This leaves some 10,000 tonnes of material to be brought from Earth, which is a major undertaking but not necessarily impossible.

The scale of the enterprise in relative terms is not essentially different from the circumnavigations of the globe in the Middle Ages and the great colonisations of the seventeenth and eighteenth centuries. It is technically feasible given a commitment similar to the Moon programme of the 1960s. To me the main problem seems to be a psychological one. Who will decide who should go and colonise the space cylinders? Will no one want to go or will everyone want to quit polluted Earth, causing a waiting list and divisive decisions on priorities? I suspect that the former is likely to be nearer the truth, at least initially. But this may not be a bad thing. In the first instance the prototype will hardly be habitable other than by committed specialists; 'real' colonies will follow from the experience gained.

O'Neill's article in 1974 projected that the prototype station could be under way by 1988. That time has now arrived and there is no sign of action so far. All that we have to stimulate high technology is the so-called 'big science' – which Britain at least is beginning to question its ability to afford – and military extravaganzas such as the Star Wars strategic defence initiative.

Surely we can do better than this. We *have* to!

Future Life

The Universe has existed for 20 billion years so far. The human species has developed in just 1 million years, a trifling one twenty-thousandth part of the whole span. If the time-span from the Big Bang to the present is likened to a cosmic year, then human life developed during the last half an hour on New Year's Eve.

The whole genus, of which we are the pinnacle, began only during the last afternoon; the first primates arrived only during the last day; the first mammals emerged during the last week; the first insects began only during the final fortnight. No life at all on Earth through the whole spring, summer and autumn. The first single-celled organisms only began to emerge in November and then in the last days of the winter an explosion of hierarchies and forms.

When we look at the Universe in this compacted time-frame we see dramatically the exceedingly rapid emergence of complicated biological structures. If we now look to the future we might just be able to imagine what possibilities there are for life forms yet to come.

The Universe is evolving, expanding and cooling. If it continues to do so then various disasters begin to occur after 10^{11} years (see page 180 above). Eventually matter itself may be expected to erode away, probably on the time-scale of 10^{30} years. Suppose that the Universe has a life of that length; where are we now?

If we liken the entire life of the Universe to a human scale of 100 years, then the life to the present corresponds to a fraction of a second after conception.

Suppose that all we knew of life was the existence of the newly formed egg. Would we imagine the foetus, baby, adult? Surely our imagination would not generalise so far. *Homo sapiens*, in the cosmic scale, is like that barely formed egg. Assuming that we get through the next few nuclear decades, what forms will develop aeons from now? On such a scale, and bearing in mind how rapid the evolution has been so far, I think we can assume that whatever is possible in principle will occur in practice. Where does that lead us?

Even to think about this in any serious fashion presupposes that we know that 'life', consciousness, actually is. If it is no more than a complicated collection of molecules then it may become possible to transmute life forms in the future.

Imagine a future where biological science and technology are sufficiently advanced that a complete map could be made of you at the molecular,or even atomic level. Your exact constitution in terms of atoms of carbon, calcium, oxygen and so forth would be known. The way that they fit together, everything that defines 'you' in terms of the organic molecules comprising your brain, all of these would be listed in some super computer.

So there is a blueprint for building 'you'.

Now we need an intricate machine that can go to a chemical bank and select a few thousand billion billion carbon atoms, a similar number of nitrogen and so forth and assemble them according to the blueprint. In fact it might be less complicated than this, in that banks would store pre-assembled molecules rather than the constituent atoms. Eventually the machine will fit all of these together in exactly the same combinations as exist in the real you. Would this new collection of molecules think it was you? If consciousness is entirely a result of chemical structures, then presumably the collection would, at that instant, have the same memories and thoughts as you had when the blueprint was made.

If this is true, we could vaporise a person into their constituent molecules and reassemble them at a later stage. If we combine the jigsaw pieces faithfully as they were immediately prior to vaporisation then we will have the same jigsaw, the same person existing again – presumably unaware that anything untoward has taken place.

Once *Homo sapiens* is able to store and reconstruct the essential biological molecules it will be freed from the human body form. It can adopt whatever form best suits it. In an ageing, cooling Universe, as the stars die out and energy supplies are exhausted, the surviving life forms will be those that can best adapt themselves to low energy demands.

Freeman Dyson has thought a lot about the realities of this, and has shown that there is nothing that violates any known laws of Nature. Nature may well allow it in principle; given time-scales billions of epochs longer than we have existed, is it unreasonable to suppose that the technology will turn it into a practicality? As Dyson says, 'We could not imagine the architecture of a living cell of protoplasma if we had not seen one'. In similar vein, we could not imagine the new born baby if all that we have seen was the undivided egg. Nor can we foresee the forms life will take on in another billion billion years.

This brings us back to the question 'What is life?'. I don't mean what gives life its 'vital force'; rather, what *defines* life? After all, we recognise whether things are 'alive' so we have some intuitive feeling for what life is. More relevant for our contemplation of the long-term future is the survival of *intelligent* life.

Computer scientists are currently developing so-called 'intelligent' machines. To know whether a machine is intelligent or not we subject it to the 'Turing test'. (Alan Turing was a mathematical genius who died in 1954 when only 41 years old. He played a leading role in the development of modern computing and was the inventor of the test named after him.)

Imagine that you are sitting at a terminal in one room and use it to communicate with another terminal which is operated by an unseen individual in another room. You pose questions at your terminal and receive responses from the next room. Can you tell from the responses whether it is a real person – an intelligent being – or a machine in the other room?

If you play chess with it and if you are a reasonable player you might quickly decide it is a machine – or a poor player! But now that chess-playing computers operate at a sophisticated level, it may prove hard to decide. If it is impossible to decide by interrogation whether you are interacting with a computer or a person, then the other room contains an intelligent source. A machine that passes this test is said to be intelligent.

These ideas from computing can be turned around and taken over into the realm of biological systems; we can imagine a living creature as a type of intelligent computer. In computing jargon, the computer is the hardware and the program is the software. In the biological example, a human is a *program* that has been designed to run on a particular type of hardware – namely the human body. The data are encoded in a special type of store, namely DNA molecules and nerve cells. 'Life' then becomes equivalent to 'information processing'.

If we are no more than an intelligent program, capable of passing the Turing test, then it is possible that this program could run on a different type of hardware. To some extent this is a technical way of stating what we have already discussed, namely, whether we could reassemble the essential biological molecules into new, non-humanoid forms? However, the Turing way of posing the problem goes far beyond this. Implicitly it raises the question of whether we even need the carbon atoms and the molecules at all. Our culture and life itself may be able to continue, in the sense of information processing, using quite different forms of matter from those we are used to. Indeed, if carbon and the constituent protons erode away, as Chapter 11 discussed, there will come a time when the Universe no longer contains these essential ingredients. So, even if our descendants invent the necessary technology, there is the question of whether 'we' will be able to survive *in principle*.

First, to build the system we will need to exploit whatever matter there is remaining in the dying universe. There must also be enough energy around to run it. If an unlimited amount of information can be processed in the future, then 'life' can exist for ever.

In an expanding cooling universe it may indeed be possible to satisfy all the constraints and survive.

The unit of computer storage is called the 'bit' (the term is shortened from 'binary digit', and is applied to the computer's ability to store the result of a choice between two alternatives). The processing of information involves manipulating 'bits per second'. The laws of physics, in particular thermodynamics, limit the rate of information processing. At room temperature it is impossible to exceed 10^{21} bits for every watt of power in a second. Modern computers are well within this constraint – an IBM personal computer has 10^8 and a CRAY supercomputer still only 10^{10} bits per second per watt.

Now consider the human machine. A human thought, a moment of consciousness, lasts about 1 second. Each of us dissipates about 200 watts of power at room temperature (similar to a light bulb). So we are at most 10^{23} bits (the 10^{21} bits multiplied by 200 watts). To maintain an intelligent society of 1 billion individuals, some 10^{33} bits are required. That is the target that we have somehow to meet.

The act of processing information generates waste heat. More is metabolised the more bits are involved and the higher the temperature. We live at room temperature; future life forms may be able to choose what temperature is optimal for them. The lower the temperature, the less waste heat there will be. But you mustn't get too cold; you must be above the ambient temperature of the universal microwave background radiation (currently −270 degrees Centigrade) and you must be able to radiate away the waste heat generated by metabolism.

The most likely long-term surviving particles of matter will be electrons. Electrons radiate electromagnetic energy – light, heat and radio waves. Nature limits the rate at which this can occur. So for life to continue the creature must metabolise less waste heat than the maximum it can radiate away. Hibernation can help it keep within bounds. Life may metabolise intermittently while continuing to radiate away waste heat during hibernation.

A creature's subjective awareness of time depends upon its metabolic rate. This subjective time may have little relation to 'real' time. Freeman Dyson has analysed the effect of this hibernation and concludes that, although biological clocks would be slowing down, going on and off as the Universe expands and cools, subjective time may go on forever. Moreover, a *finite* amount of energy is needed for indefinite survival. Between now and the end of never, a society with the complexity of the human species will use only as much energy as the Sun radiates in 8 hours. So the problem of energy reserves is trifling. The reserves in a galaxy can sustain a society a billion billion times more complex than ours.

Thus in principle there is nothing yet known in physics that prevents this. The technology and the architectures of life forms are less certain. One feature that will be needed is memory – to be immortal with a finite memory would be undesirable. Dyson has ideas on how we might store information by ordering the remaining electrons – little magnets oriented up or down – like a binary code.

A philosopher wrote that there is one certain truth, namely that life has no purpose. Yet there is a great psychological urge to seek purpose in it all. Preservation of the human species and development of awareness and culture are threads that link the generations, hence the interest in extinction: you and I will surely die some day but others will continue in the human relay race. The possibility that the race will end does concern us.

When I think of what has happened in the last 50 years I wonder whether we can survive the next 50 let alone worry about natural extinction. Scientists are evaluating the prospects for surviving a nuclear winter. Now we suddenly discover that aerosol sprays are ripping a hole in the protective ozone layer over the poles; this will let in solar radiation that could be lethal. These are man-made problems amenable to immediate political solutions.

Then there are threats that we may or may not be able to deal with. We are in an ecosystem along with a complex microbiology. Viruses and other disease-spreading organisms nutate. Nature always keeping one step ahead of our efforts. A cracked container in a germ warfare laboratory has been a favourite life-threatening scene in some novels, but Nature could do it itself – I know of no general principle saying that human science can necessarily combat every mutation, at least in the time-scale required.

For example, when I started writing this book AIDS was little known by the general public; now it is perceived as a potential threat to the species. In an extreme, the human race could contract, dominated by ascetics, unmarried virgins and a percentage of monogamous pairings. The population might fall by a fraction of a per cent or by a large amount. In the latter case there would be a profound change in our socio-economic structure until the virus died out with no hosts to propagate it. Humans would reflower, however.

The chances of a natural catastrophe in the short term are small and, though guaranteed in the long term, there is plenty that we can do to prepare. In the great extinctions of the past, many life forms survived and next time around we should be intelligent enough to make it with the ants.

The most dangerous things in the Universe may well be humans. We are currently in a critical period where we have to elevate our moral stature to match our rapidly increasing scientific and technological development. If ever we meet advanced creatures from other worlds it would be encouraging to know that those societies have proved that it is possible to do this. If there is a message in this book it is this: we are not omnipotent. We are much more vulnerable to external circumstances than we would like to admit. If we can humble ourselves to realise that, then maybe we can begin to face our self-made dangers with more responsibility and urgency. If we don't, then we will wake up one morning and find we're not here!

Suggestions for Further Reading

I have listed here some source material (with an arbitrary restriction to being post-1982 except in a few classic cases) together with articles that may help you probe further into some of the themes in this book. The list is by no means complete or authoritative; it simply contains articles that I found useful at some stage or other. In many cases they contain references to earlier works which should enable you to research topics in depth. By its very nature, the frontier of research is often controversial and I have chosen a personal perspective. Many of the issues are still being argued over and these articles will help give a flavour of some of the disputes.

Impacts, Meteorites and Comets

R. Harrington and T. van Fladern, *Icarus*, vol. 39 (1979), p. 131.

R. Grieve and P. Robertson, 'Earth craters', *Icarus*, vol. 38 (1979), p. 212.

W. Napier and V. Clube, 'A theory of terrestrial catastrophism', *Nature*, vol. 282 (1979), p. 455.

R. Kerr, 'Impact looks real, the catastrophe smaller', *Science*, November 1981, p. 896.

D. Hughes, 'The first meteorite stream', *Nature*, September 1982, p. 14.

R. Ganapathy, 'A major meteorite impact on Earth 34 million years ago: Implications for Eocene extinctions', *Science*, May 1982, p. 885.

'Close encounters in Space', *Sky and Telescope*, June 1982, p. 570.

D. J. Michels, 'Observation of a Comet on collision course with the Sun', *Science*, vol. 215 (1982), p. 1097.

R. Ganapathy, 'Tungusku: Discovery of meteorite debris . . . ', *Science*, June 1983, p. 1158.

R. Grieve, 'Impact craters shape planet surfaces', *New Scientist*, November 1983, p. 517.

'Geological rhythms and cometary impacts', *Science*, December 1984, p. 1427.

D. Hughes, 'Meteorites from Mars', *Nature*, October 1984, p. 411.

R. Knacke, 'Cosmic dust and the Comet connection', *Sky and Telescope*, September 1984, p. 206.

I. Halliday *et al.*, 'Frequency of meteorite falls on Earth', *Science*, March 1984, p. 1405.

D. Steel and W. Baggaley, 'Collisions in the solar system – Impacts of the Apollo asteroids upon the terrestrial planets', *Monthly Notes of the Royal Astronomical Society*, vol. 212 (1985), p. 817.

E. Marshal, 'Space junk grows with weapons tests', *Science*, October 1985, p. 424.

'Voyager 2 at Uranus', *Sky and Telescope*, November 1985, p. 42.

M. Waldorp, 'Voyage to a blue planet', *Science*, February 1986, p. 916.

A. P. Boss, 'The origin of the Moon', *Science*, January 1986, p. 341.

'Io spirals towards Jupiter', *New Scientist*, January 1986, p. 33.

'Giotto finds a big black snowball at Halley', *Science*, March 1986, p. 1502.

'Are cometary nuclei primordial rubble piles?' *Nature*, March 1986, p. 243.

Extinctions

L. Spencer, *Mineralogy Magazine*, vol. 295 (1939), p. 425.

J. Laurence Kulp, 'The geological time scale', *Science*, vol. 133 (1961), p. 1105.

S. Durrani, *Physics of the Earth and Planets*, vol. 4 (1971), p. 251.

N. Snelling, 'Measurement of the geological time scale,' Talk at British Association of Science, 1987; editor of *Chronology of the*

Geological Record, Geological Society of London, Memoir No. 10 (published by Blackwell, Oxford, 1985).

H. Urey, *Nature*, vol. 242 March (1973), p. 32.

A. W. Alvarez *et al.* 'Extraterrestrial cause for the Cretaceous – Tertiary extinction', *Science*, June 1980, p. 1095.

R. Ganapathy, 'A major meteorite impact on Earth 65 million years ago: Evidence from the Cretaceous–Tertiary boundary clay', *Science*, August 1980, p. 921.

R. Ganapathy, 'A major meteorite impact on Earth 34 million years ago: Implications for Eocene extinctions', *Science*, May 1982, p. 885; for an opposing opinion see G. Keller *et al.*, *Science*, vol. 221 (1983), p. 150.

P. J. Smith, 'The origin of tektites – settled at last?' *Nature*, November 1982, p. 217.

'Extinctions and ice ages – are comets to blame?' *New Scientist*, June 1982, p. 703.

C. Officer and C. Drake, 'The Cretaceous Tertiary transitional', *Science*, March 1983, p.1383.

'Extinctions by catastrophe?' Five articles in *Nature*, April 1984, pp. 709–20, and commentary p. 685.

'Periodic impacts and extinctions', *Science*, March 1984, p. 1277 (Report of a workshop on comet impacts and their effect on evolution).

'Mass extinctions in the ocean', *Scientific American*, June 1984, p. 46.

'Geological rhythms and cometary impacts', *Science*, December 1984, p. 1427.

'Ammonoids and extinctions', *Nature*, January 1985, p. 12 and pp. 17–22.

The dinosaur controversy: *Nature*, June 1985, pp. 627 and 659; *Science*, March 1985, p. 1161; *New Scientist*, November 1984, pp. 9 and 30.

'Extinctions ARE periodic', *New Scientist*, March 1986, p. 27.

R. E. Sloan *et al.*, 'Gradual dinosaur extinction', *Science*, May 1986, p. 629.

C. B. Officer *et al.*, 'Cretaceous and paroxysmal Cretaceous–Tertiary extinctions', *Nature*, vol. 326 (1987), p. 143.

The Sun

J. Eddy, 'The Maunder Minimum', *Science*, vol. 192 (1976), p. 1189.
F. Close, 'Is the Sun still shining?', *Nature*, vol. 284 (1980).
J. Gribbin, *The Strangest Star*, Fontana, 1980.
'100 to 200 year solar periodicities', *Nature*, September 1982, p. 14.
S. Sofia *et al.*, 'Solar radius change between 1925 and 1979', *Nature*, August 1983, p. 520.
J. Parkinson, 'New measurement of the solar diameter', *Nature*, August 1983, p. 518.
'Chasing the missing solar neutrinos', *New Scientist*, January 1984, p. 20.
R. A. Kerr, 'The Sun is fading', *Science*, January 1986, p. 339.
Hans Bethe on solar neutrinos, *Nature*, April 1986, p. 677.
G. Williams, 'The solar cycle in Precambrian time', *Scientific American*, 1986, p. 80.
R. Bracewell, 'Simulating the sunspot cycle', *Nature*, vol. 323, (1986), p. 516.
F. Paresce and S. Bowyer, 'The Sun and the interstellar medium, *Scientific American*, September 1986, p. 89.
'600 million years of solar cycles', *New Scientist*, October 1986, p. 29.
W. Haxton, 'The solar neutrino problem', *Comments on Nuclear and Particle Physics*, vol. XVI (1986), p. 95.
J. Bahcall, G. Field and W. Press, 'Is solar neutrino capture rate correlated with sunspot number?', Princeton University report, 1987.

The Galaxy

F. Hoyle and R. Lyttleton, 'The effect of interstellar matter on climate variation', *Proceedings of the Cambridge Philosophical Society*, vol. 35 (1939), p. 405.
W. H. McCrea, 'Ice-ages and the galaxy', *Nature*, vol. 255 (1975), p. 607.
L. Blitz *et al.*, 'The new Milky Way', *Science*, June 1983, p. 1233.
'The Earth's orbit and ice-ages', *Scientific American*, February 1984, p. 42.

'A black hole at the galatic centre', *Nature*, vol. 315 (1985), p. 93.
'The galactic centre: Is it a massive black hole?' *Nature*, September 1986, p. 1394.

Stars and Supernovae

E. Norman, 'Neutrino astronomy: a new window on the universe', *Sky and Telescope*, August 1985, p. 101.
D. Patterson, 'A supernova trigger for the solar system', *New Scientist*, May 1978, p. 361.
A. Burrows and T. Mazurek, 'Signatures of stellar collapse in electron-type neutrinos', *Nature*, January 1983, p. 315.
M. G. Edmunds, 'Upper limit on the mass of a star (The decline of a superstar)', *Nature*, December 1983, p. 741.
'Dead Stars', *Nature*, March 1984, p. 142.
S. Chandrasekhar, 'Stars: their evolution and stability' (Nobel Lecture), *Science*, November 1984, p. 497.
'Supernova sets limit on neutrino mass', *New Scientist*, March 1987, p. 24.

The Nature of Matter

F. Close, *The Cosmic Onion*, Heinemann Educational, 1983 (American Institute of Physics, 1987).
F. Close, M. Marten and C. Sutton, *The Particle Explosion*, Oxford University Press, 1987.
C. Sutton, *Building the Universe*, Blackwell, 1985.
M. Turner and F. Wilczek, *Nature*, vol. 298 (1982), p. 633.
M. Rees and P. Hut, *Nature*, vol. 302 (1983), p. 508.
E. Farhi and A. Guth, *Physics Letters*, vol. 183B (1987), p. 149.
E. Witten, 'Strange matter', *Physical Review*, vol. D30 (1984), p. 272.
M. Green, 'Superstrings', *Scientific American*, September 1986, p. 44.

Dark Matter and the Missing Mass

'Dark matter in spiral galaxies', *Scientific American*, June 1983, p. 88.

V. Rubin, 'The rotation of spiral galaxies', *Science*, June 1983, p. 1339.

L. Blitz *et al.*, 'The new Milky Way', *Science*, June 1983, p. 1233.

J. Silk, 'The black cloud', *Nature*, September 1983, p. 388.

'New light on dark matter', *Science*, June 1984, p. 971 (Report on a workshop at Fermi Laboratory, Chicago).

P. Hut and S. White, 'Can a neutrino dominated universe be rejected?' *Nature*, August 1984, p. 637.

'Candidates for cold dark matter', *Nature*, October 1984, p. 517.

'Tracking down the missing mass', *New Scientist*, January 1986, pp. 32, 37–40.

J. Bahcall and S. Casertano, *Astrophysical Journal*, vol. 293 (1985), p. 1–7.

'New clues to galaxy formation', *New Scientist*, January 1986, p. 34.

M. Rees, *Monthly Notes of the Royal Astronomical Society*, vol. 218 (1986), p. 25.

M. Waldrop, 'In search of dark matter', *Science*, September 1986, p. 1386; October 1986, p. 152.

The Future

F. J. Dyson, 'Time without end', *Reviews of Modern Physics*, vol. 51 (1979), p. 447.

G. K. O'Neill, 'The colonisation of space', *Physics Today*, September 1974.

J. Barrow and F. Tipler, 'Eternity is unstable', *Nature*, vol. 276 (1978), p. 453.

'Thermodynamics of a closed universe', *Nature*, March 1984, p. 319.

Glossary

alpha particle A nucleus of helium comprising two *protons* and two *neutrons* tightly bound to one another. May be emitted in the radioactive decay of a *nucleus (fission)*.

antimatter For every variety of fundamental particle there exists an antiparticle with the same mass but with the opposite sign of the electrical charge. When a particle and its antiparticle meet they can mutually annihilate and liberate energy.

asteroid Small rocky objects orbiting the Sun. Most orbit between Mars and Jupiter but several have trajectories that cross our own. Remnants of *comets* and the debris from collisions between a handful of small bodies that condensed between Mars and Jupiter during the formation of the solar system. Also known as minor planets or planetoids.

atom System of *electrons* orbiting a *nucleus*. Smallest piece of an element that can still be identified as that element.

aurora Display of diffuse light seen high in the atmosphere mainly in extreme polar regions. Caused by charged particles trapped in the Earth's magnetic field.

beta decay Decay of a radioactive *nucleus* with production of an electron *(beta particle)*. The underlying process is the transmutation of a *neutron* into a *proton* with *electron* and *neutrino* emission. This process is controlled by the *weak interaction* and was its first known manifestation.

beta particle *Electron* emitted in radioactive decay of a *nucleus* (*beta decay*).

Big Bang The galaxies are receding from one another: the Universe is expanding. The Big Bang theory proposes that this expansion began 10–20 billion years ago when the Universe was in a state of enormous density.

black hole A region where gravity is so powerful that light cannot escape. Hypothesised to be formed when some stars collapse. The gravitational influence of the collapsed star can be felt but no information escapes from within the black hole.

comets Members of solar system travelling around the Sun on elongated orbits. Thought by many astrophysicists to be akin to mile-sized dirty snowballs. So tenuous that the *solar wind* blows material from them and gives rise to long tails of gas and dust that is their familiar manifestation in the sky.

cosmic rays High-speed particles and nuclei coming from outer space.

deuteron *Nucleus* comprising one *proton* and one *neutron*. Sometimes called heavy hydrogen. Produced at an intermediate stage of stellar *fusion*.

electromagnetic force One of the fundamental forces of Nature. Attraction of oppositely charged particles traps negatively charged *electrons* around a positively charged *nucleus* in *atoms*.

electromagnetic radiation Energy emitted by electrically charged objects. Familiar examples include light, radio waves, microwaves, X-rays and *gamma rays*.

electron Negatively charged elementary particle constituent of *atoms*. Carrier of electricity through wires.

electronvolt (eV) Unit of energy. 1 eV is the energy gained when an *electron* is accelerated by a potential of one volt.

fission Break-up of a large *nucleus* into smaller ones.

fusion Combination of small nuclei into larger ones.

gamma ray Photon. Very high-energy *electromagnetic radiation*.

ion *Atom* carrying electrical charge due to its being stripped of *electrons* (positive ion) or having an excess of electrons (negative ion).

meteor The streak of light seen in a clear night sky when a small particle of interplanetary dust or stone burns itself out in the Earth's upper atmosphere.

meteorite A *meteor* that makes landfall; remnants of *meteoroids* found on the ground.

meteoroid Collective term applied to meteoric material in the solar system.

molecule Cluster of *atoms*.

nebula Gas or a collection of stars that appear as a bright mist in the clear night sky.

neutrino Electrically neutral elementary particle, with little or no mass. Most abundant particles in the universe. Only takes part in *weak interactions*; produced in radioactive processes and as a by-product in stellar *fusion* and *supernovae*.

neutron Electrically uncharged partner of the *proton* in atomic nuclei.

nucleus Dense centre of *atoms* built from *neutrons* and *protons*. The latter give the nucleus a positive electrical charge by which *electrons* are attracted and atoms formed.

photon Bundle of *electromagnetic radiation* and carrier of *electromagnetic forces*.

plasma A gas of *ions* and *electrons* moving freely. A plasma is in effect an ionised gas.

positron Antiparticle of *electron*. Carries positive electrical charge. Stable until it meets an electron, at which point the pair may mutually annihilate into *electromagnetic radiation*.

proton Positively charged constituent of the *nucleus* that give it electrical charge. Built from three *quarks*.

protostar An early stage in the formation of a star. It has fragmented from a gas cloud and started to collapse but nuclear reactions have not yet begun.

quark Believed to be one of the fundamental constituents of matter. Clusters of quarks comprise the *neutrons* and *protons* in the atomic *nucleus*.

quasar A contraction of the name Quasi-Stellar Object. A compact extragalactic object that looks like a point of light but emits more energy than hundreds of galaxies combined.

They are among the most distant objects that we have observed in the Universe.

radioactivity Spontaneous decay and transmutation of *nuclei* with emission of particles comprising alpha, beta or gamma radiation.

shooting star Another word for a *meteor*.

solar flare Sudden short-lived brightening of the Sun's visible surface caused by an explosive release of energy in the form of particles and radiation.

solar wind Flow of electrically charged particles, mainly *electrons* and *protons*, from the Sun into the solar system.

strange matter Hypothesised form of matter whose atomic *nuclei* contain a substantial percentage of *strange particles* in addition to the *quarks* that comprise the *neutrons* and *protons* of normal nuclei.

strange particles Variety of particles that are not stable under normal conditions on Earth but may stabilise within *strange matter*. They have been seen fleetingly in *cosmic rays* and produced in particle accelerators.

sunspots Comparitively dark regions on solar surface, their number varying on roughly an 11-year cycle (the sunspot cycle).

supernova Sudden appearance or brightening of a star caused by an explosion that blows off most of the star's outer material accompanied by collapse of the remainder into a dense ball of *neutrons* (neutron star) or *black hole*.

weak interaction (force) One of the fundamental forces of Nature. Most famous manifestation is in *beta decay*, also involved in some radioactive decays of *nuclei* and *neutrino* interactions.

WIMPS Acronym for Weakly Interacting Massive Particles, hypothesised particles that are more massive than *protons* and which take part in the *weak interaction*. They may have been formed soon after the *Big Bang* and still occur in the centre of stars like the Sun. If so, they may affect the internal fuel cycle in the Sun.

Index

E

F

G

H

I

J

K

L

M

Q

R

S